MARXISM AND EDUCATION

This series assumes the ongoing relevance of Marx's contributions to critical social analysis and aims to encourage continuation of the development of the legacy of Marxist traditions in and for education. The remit for the substantive focus of scholarship and analysis appearing in the series extends from the global to the local in relation to dynamics of capitalism and encompasses historical and contemporary developments in political economy of education as well as forms of critique and resistances to capitalist social relations. The series announces a new beginning and proceeds in a spirit of openness and dialogue within and between Marxism and education, and between Marxism and its various critics. The essential feature of the work of the series is that Marxism and Marxist frameworks are to be taken seriously, not as formulaic knowledge and unassailable methodology but critically as inspirational resources for renewal of research and understanding, and as support for action in and upon structures and processes of education and their relations to society. The series is dedicated to the realization of positive human potentialities as education and thus, with Marx, to our education as educators.

Renewing Dialogues in Marxism and Education: Openings
Edited by Glenn Rikowski, Helen Raduntz, and Anthony Green

Critical Race Theory and Education: A Marxist Response
Mike Cole

Critical Race Theory and Education

A Marxist Response

Mike Cole

palgrave
macmillan

CRITICAL RACE THEORY AND EDUCATION
Copyright © Mike Cole, 2009.

First published in 2009 by
PALGRAVE MACMILLAN®
in the United States—a division of St. Martin's Press LLC,
175 Fifth Avenue, New York, NY 10010.

Where this book is distributed in the UK, Europe and the rest of the world,
this is by Palgrave Macmillan, a division of Macmillan Publishers Limited,
registered in England, company number 785998, of Houndmills,
Basingstoke, Hampshire RG21 6XS.

Palgrave Macmillan is the global academic imprint of the above companies
and has companies and representatives throughout the world.

Palgrave® and Macmillan® are registered trademarks in the United States,
the United Kingdom, Europe and other countries.

ISBN-13: 978–0–230–60845–0 (hardcover)
ISBN-10: 0–230–60845–0 (hardcover)
ISBN-13: 978–0–230–61335–5 (paperback)
ISBN-10: 0–230–61335–7 (paperback)

Library of Congress Cataloging-in-Publication Data is available from the
Library of Congress.

A catalogue record of the book is available from the British Library.

Design by Newgen Imaging Systems (P) Ltd., Chennai, India.

First edition: March 2009

10 9 8 7 6 5 4 3 2 1

Printed in the United States of America.

To Susi Knopf with thanks and with love

The problem of the Twentieth Century is the problem of the color-line.
                                                                W.E.B. Du Bois

The history of all hitherto existing society is the history of class struggles.
                                                                Karl Marx

Contents

Series Editor's Foreword

This book is a timely and very welcome contribution to the spirit and substance of the Marxism and Education Series in terms of open dialogue and critique. Mike Cole's central focus is an appraisal of Critical Race Theory (CRT) whose main radical theme is that 'race' is the fundamental issue, 'race trumps class', as some CRT analysts are fond of asserting. Importantly, dialogue and critique figure centrally where we engage with exploitation and oppression, in this case racism and its educational ramifications most broadly conceived. This is notably so when articulating 'race' and class as they are played out in and for capitalist relations of production. Seen in this way, some CRT formulations provide a major challenge to Marxism. Arguably the CRT approach sets up priorities that foreshorten the need to *articulate* racism with capitalist relations, or indeed in extremis, deny the necessity to elaborate critique of capitalist political economy of education at all. In this context Cole takes up these challenges and provides a comprehensive review of CRT transatlantic literature, including an account of its origins in a radical professional movement within U.S. legal discourses and demonstrates its recent leading UK developments. In doing so, his point is to elaborate a multifaceted and nuanced analysis. From within Cole's Marxist position it is therefore very important to recognize the coherence and cogency of CRT. In this light he examines the emergent theoretical specificity and political implications for informing radical and progressive practices and so contributes to the development of Marxist/CRT dialogue.

While Mike Cole's approach clearly recognizes the strengths of CRT, building immanent critique of CRT for theory and practice is important, too. His analysis signals points where CRT's dialectical negating the negative of racism fails, and where he offers his own more robust Marxist ideology/ critique. Thus, CRT may be argued to be (inadvertently perhaps) playing into the hands of some of the oppressive forces it aims to critique. Here Cole's analysis elaborates the ways in which CRT adopts a narrowly focused concept of 'white supremacy' as its analytically irreducible point of social antagonism. His analytical model provides the basis for comradely critique of CRT articulated around his theory of racialization (elaborated in relation to the work of Robert Miles) and through it he aims to demonstrate how these racist forms have ideological effects of buttressing capitalism. The point is that under capitalism whites and blacks (if we are to work for the moment with such a simplified social dichotomy) live their lives as part of an emergent class system in which it is not the case that all whites are dominant, far from it. In Marxist analysis, oppressed and poor blacks and whites have too much in common in class terms for such things to be diverted and cancelled out by surface appearances which undermine the potential for social and political cooperative relations. CRT and Marxism grow uneasy with one another and tend to part company over such critical issues. For Marxism and education the main problem is to draw attention to the significance of complex appreciations of the articulation of 'race' and class relations. Here the potential for progressive cooperation between Marxism and CRT is strained, and may break down in struggles over educational inequity, for instance. The general point is that for Marxism antiracism in and around education must articulate with a contextually specific structural analysis of capitalism to understand the possibilities of progressive forces cohering for articulating class both *in and for itself.* In this respect those cultural forms in which there is little or no recognition of non-color-coded racism can be just as disruptive of progressive alliances as can ignorance, misrecognition and rejection of the relevance of skin color varieties of racist ideology. What is at stake here is the possibility of dialogue with Marxism to enhance CRT's strengths and building progressive alliances in struggle.

As a contribution to the Series, Mike Cole's book is welcomed because it is both relentlessly political in its class analysis and is educational in its form and intent, as well as being engaging as a strong personal statement. It is also sharply controversial and necessarily provocative. In each respect he again expresses one of the foundational historical materialist points that the philosophers have up till now been intent upon and contented to understand the world, the crucial point however, is to change it! As a Marxist educator, Cole reaffirms themes he has elaborated elsewhere (Cole, 2008f) concerning a variety of familiar critical points raised against Marxism to which he offers his own series of rebuttals as part of the struggles in and for education. And he provides many practical ideas for antiracism at the classroom level, as well. Finally, the book is enhanced by its attention to contemporary manifestations of struggles, most notably in his account of antiracist politics in the Bolivarian Republic of Venezuela.

This book is to be understood as a complex and dynamic totality. As such it is noteworthy that a foundational characteristic of CRT is that it is concerned to recognize the authenticity of *voice* in political and cultural complexity and specificity. On this count Cole's analysis reminds us that this is of vital concern for what is socially constitutive but that it should not be deployed to naively contradict or even disperse materialist analysis in shallow empiricist fashion. Appearances may deceive. The issue is to analyze and socially account for what is happening, including how things appear to be the case. If global capitalism is to be seriously challenged and transcended, it is a crucial dimension of any materialist and realist analysis that is likely to be able to contribute to building progressive alliances at local, regional, national and trans-national levels, that cultural forms and experience, histories and identity narratives must be recognized as real terrains, as media and as resources for struggle. What is at stake then is building common grounds of class identity *for itself* in and through such complexity. The methodological issue about *being and consciousness* is not a matter of *either/or* so far as cultural and material forms are concerned. It is about multilayered and complex articulation, always already indicating *both/and* requirements in these respects for concrete analysis and action. Mike Cole's book marks a further contribution to the continuing development of the legacy of Marx and the Marxist traditions in, as and for education through its robust and engaging critical working of theory and political practice.

ANTHONY GREEN
August 2008

Acknowledgments

A number of people have helped me with this book in various ways, from suggesting sources and references to reading and commenting on whole chunks of text. I am indebted to a number of people for their help: Fabrice Bensimon; Derrick Bell; Terrell Carver; Angelina Castagno; Namita Chakrabarty; Antonia Darder; Adrienne Dixson; Jean Ducange; Gavin Hayes; Tom Hickey; Dave Hill; Susi Knopf; Stathis Kouvelakis; Gloria Ladson-Billings; Sheila Macrine; Alpesh Maisuria; Brian Matthews; Charles Mills; Andy Pilkington; John Preston; Rodolfo Torres; Gabrielle Tree; Rebecca Tsosie; Paul Warmington; and Richard Woolley. Special thanks must go to Anthony Green, the series editor, and to Richard Delgado for their encouragement, support and instantaneous critical and insightful responses to various drafts of this book. The above named acquaintances, friends and comrades will receive this book with varying degrees of response. It goes without saying that none of them should be held responsible for any inadequacies therein.

Introduction

Some Early Personal Experiences of Racist Britain

Growing up in Bristol, a key pivot of the slave trade in the eighteenth and early nineteenth centuries, my first recollection of the manifestations of racism was encapsulated in a childhood saying. The saying, obviously directed at girls, was that 'if you step on the lines between the stones on the pavement, when you grow up you'll marry a black man'. I also recall going for a walk with my grandmother, and her pointing out with surprise, 'look there's a black man!'.[1]

A little later in my childhood, I think it the late 1950s, as immigration was increasing, I remember being told (although I do not know whether it was true) that a well-known chain store had a policy of not employing black labor. As a young child, the racist norms of the society must already have affected me. Having seen black women working in Woolworths, I remember thinking that this other popular chain store must be in some way superior because of its operation of this 'color bar'. About the same time I remember my cousin remarking to me disapprovingly that 'Jamaicans' spent more money on their cars than on their homes. At the time it was commonly accepted that black people ate Kit E Kat (a popular cat food) because they knew no different.

At school, in the early 1960s if I recall correctly, I found myself being driven home from school by a medical doctor, who happened to be the father of a friend. As we passed near St. Pauls, an area with a long-standing black population, he told me that that was where the 'jungle bunnies' live.

One of my best friends at college in London in the late 1960s (where I studied for and failed a sociology degree) was a Trinidadian of Asian[2] heritage. Our friendship continued as I embarked on a teacher education course at another college. We used to meet up regularly to drink beer and eat curry in a couple of rooms in which I lived in Kew Gardens, in the southwest of London. I remember vividly the reaction of the landlady, on discovering his presence: 'it's not right having a black man in a white house'. When challenged, she responded by stating, 'it's not *so* bad him being in your kitchen, but I do object to cleaning the toilet after him' (we had our own kitchen, but shared the toilet with the landlady).

By the 1970s, I was teaching in a primary school in Ladbroke Grove in west London, determined that I would use my role as a teacher to challenge racism, and all the other inequalities that, as a Marxist educator, I had decided was one of my main goals. The opening remarks about my final year primary class (ten- to eleven-year-olds) from the Deputy Head before my first meeting with the class was 'you won't get anything out of them'. Determined to prove her wrong, I decided to change the order of things somewhat. During morning assembly, the (overwhelmingly) African-Caribbean children were forced by the Head to sit cross-legged on the floor and to listen to western classical music, while the Deputy Head moved around the hall and coerced them into order and silence. I insisted that on the first day at the school (I was employed by the Inner London Education Authority on an enhanced salary to work temporarily in schools that were having difficulty retaining staff) that I would not teach the whole class, but would meet all the children, either individually or in pairs. At these meetings, several were surprised that we had a mutual interest in reggae music. I suggested to some of them that they bring in some records the following day. One of the first deals I negotiated with my class on that day was that, if we worked through the day, we could play some reggae in our classroom at the end of the afternoon.[3] Some months after I started teaching, six of the children's poems were published in the popular and highly respected community newspaper, *West Indian World*. Some poems were about nostalgia for Dominica, the country of origin of most of the children; others were angry tirades against the racism and class exploitation in their lives. One of the many things that sticks in my mind is a girl of Dominican origin in my class telling me that there were too many rats in her flat, and about the white man who drove round every Friday in his Rolls Royce to collect the rent.

A First Encounter with, and an Ongoing Interest in, Marxist Analyses of Racism

I first read Marx, starting with *Capital Volume 1*, nearly thirty years ago. At the same time, I became familiar with the work of the Centre for Contemporary Cultural Studies at the University of Birmingham. Headed at the time by Stuart Hall, the CCCS was publishing neo-Marxist analyses of popular culture at a breathtaking pace. Along with a number of Occasional Stenciled Papers, the Centre and its associates produced some major books (e.g., Hall and Jefferson, 1976; Centre for Contemporary Cultural Studies [CCCS], 1977, 1978; Hall et al., 1978; Clarke et al., 1979; Hall et al., 1980; CCCS, 1981). One of the Centre's books, *The Empire Strikes Back* (CCCS, 1982) dealt specifically with racism. This book, along with other Marxist analyses of racism both emanating from the CCCS and elsewhere, made me think that perhaps Marxism had most purchase in understanding the multifaceted nature of racism, both historically and contemporaneously. A few years after becoming acquainted with such analyses, I published my first Marxist critiques of racism (Cole, 1986a, 1986b) and have been using Marxist theory to try to understand racism ever since.

I am not sure when I first became aware of Critical Race Theory (CRT). However, I do remember the first critical Marxist analysis of CRT (Darder and Torres, 2004, Chapter 5) that I came across. After reading it, I began to see CRT as the latest in a long line of academic challenges to Marx and Marxism.[4] This is how Antonia Darder and Rodolfo Torres (2004, p. 117) conclude the chapter:

> any account of contemporary racism(s) and related exclusionary practices divorced from an explicit engagement with racialization and its articulation with the reproduction of capitalist relations of production is incomplete. The continued neglect by critical race theorists to treat with theoretical *specificity* the political economy of racialized class inequalities is a major limitation in an otherwise significant and important body of literature.

Since I had read and respected previous work by Darder and by Torres, I decided that I needed to read CRT in order to ascertain whether I agreed with the conclusion reached by Darder and Torres. Having read CRT, my purpose became clear: to interrogate CRT from a Marxist perspective, but also to respect some of CRT's strengths. Accordingly, Darder and Torres' critique will resonate throughout its pages. While, as will become clear in chapter 1, CRT had its origins in law, the specific focus in this volume is CRT and Education.

Outline of the Book

Before dealing with issues of educational theory, I need to set the scene. In chapter 1, therefore, I begin by briefly tracing the relationship between post-modernism, transmodernism and CRT with respect to the voices of the Other. I then examine CRT's historical origins in Critical Legal Studies (CLS) in the United States, noting how CRT was in part a response to the perception that the analyses of CLS were too class-based and underestimated the centrality of 'race' as the major form of oppression in society. I conclude with a consideration of CRT's various ethnic identity-specific varieties.

In the second chapter, I go on to critique from a Marxist perspective two of CRT's central tenets, namely the favoring of the concept of 'white suprem-acy' over racism, and the prioritizing of 'race' over class as the primary form of oppression in society. During the course of the chapter, I offer my own wide-ranging definitions of racism and the Marxist concept of racialization, arguing that these formulations are better suited in general to understanding and combating racism in the modern world, than is the CRT concept of 'white supremacy'. In this chapter, I also address the contemporaneous man-ifestations of non-color-coded racism.

In chapter 3, I examine what I perceive to be some of the strengths of CRT, namely the use of the concept of property to explain historically segregation in the United States; the all-pervasive existence of racism in the world; the importance of voice; the concept of chronicle; interest convergence theory; transposition and CRT and the law in the United States. These strengths, however, are not without limitations, and I suggest ways in which some of these strengths could be enhanced by Marxist analysis. Chapter 3 includes an appendix which features a chronicle that attempts to subvert and question the validity of the CRT concept of 'white supremacy'.

In chapter 4, I look at multicultural education in the United States and in the United Kingdom, and at the respective antiracist responses (based on Marxism) in each country. I begin by discussing three forms of reactionary multicultural education in the United States identified by Peter McLaren. I go on to analyze McLaren's advocacy, *in his postmodern phase*, of 'critical resistant multiculturalism', a form of multiculturalism favored by Critical Race Theorist Gloria Ladson-Billings. I conclude the section of the chapter on the United States by appraising McLaren's promotion, *since he returned to the Marxist problematic*, of 'revolutionary multiculturalism'.

Turning to the United Kingdom, I begin by discussing the ongoing, but now protracted, debate over the relative merits of multicultural and antiracist education. In chapter 2, I identified a threat to antiracism and in chapter 3 a threat to the acknowledgement of the existence of institutional racism, in both cases from state and other official rhetoric and policy. I conclude chapter 4 by suggesting that gains made by antiracists are further under threat from a 'hard' version of the concept of 'community cohesion', currently advocated by the UK Government.

In chapter 5, I address the relatively recent arrival of CRT in educational theorizing in the United Kingdom. In so doing, I focus on the latest book by influential United Kingdom 'race' and education theorist David Gillborn in the belief that the growing body of work by Gillborn in the field of CRT is highly likely to consolidate its presence in the United Kingdom. Specifically, I critically discuss Gillborn's views on *Marxists;* on *Marx and slavery;* on *Marx and 'species essence';* on *'White powerholders';* on *racist inequalities in the UK education system;* on *education policy;* on *ability;* on *institutional racism;* on *'model minorities';* on *whiteness;* on *conspiracy;* and on *'struggling where we are' against 'the powers that be'.*

Globalization from both CRT and Marxist perspectives is examined in chapter 6, where I also look at globalization and its relationship to the new U.S. imperialism, arguing that, as well as Marxism, some transmodern concepts but not others can aid our understanding of these processes and movements. I conclude chapter 6 with some comments on the U.S. imperialist occupation of Iraq, five years on.

I then in chapter 7 begin by addressing myself to some common objections to Marxism, followed by a Marxist response. Next I examine the alternative to capitalism of twenty-first century democratic socialism, referring to ongoing developments in the Bolivarian Republic of Venezuela. I focus on the impressive social democratic changes happening there and the possibilities of a transition to socialism. To counter CRT claims of an incompatibility between Marxism and antiracism, I conclude this chapter with a discussion of antiracism in practice in that country.

In chapter 8, I begin by looking at some areas of agreement between Marxists and certain U.S.-based Critical Race Theorists on suggestions for classroom practice, before critiquing the 'abolition of whiteness' pedagogy of a leading UK Critical Race Theorist. I then make some suggestions for practice based on Marxist theory, namely anti-imperialist antiracist education. I conclude with a call for implementing 'the last taboo', namely, to include discussions about capitalism and socialism in the school curriculum. Chapter 8 has an appendix that describes Marx's Labour Theory of Value (LTV), a theory that Marxists believe explains precisely the way in which workers are exploited under capitalism.

Finally, in the Conclusion, I begin by suggesting that CRT, while making calls for liberation and the end of oppression, in fact offers no concrete solutions for this. I reiterate that Marxism does provide a solution. Having reconfirmed that the purpose of this book is not to divide, but to unite, I look back to the struggles of Martin Luther King, in particular his attraction to socialist principles in the later part of his life. I conclude by suggesting that, now that CRT is firmly established, Critical Race Theorists might reconsider a realignment with Critical Legal Studies, thus forming a potentially fruitful partnership in the tasks that lie ahead for all progressive people.

None of the criticisms of the work of others in this book should be read as personal. Indeed, I have great respect for the various people I critique. My purpose in this book is unequivocal: namely to attempt to align us all around the project of democratic socialism for the twenty-first century, an objective socialist struggle that is fully attuned to the needs of us all. While the concerns of this book are with 'race', twenty-first century socialism must, of course, address all forms of exploitation and oppression and be fully cognizant with and address all forms of inequality.

Chapter 1

Critical Race Theory: Origins and Varieties

In this first chapter, I begin by briefly tracing the relationship between postmodernism, transmodernism and Critical Race Theory (CRT) with respect to the voices of the Other. I go on to examine CRT's historical origins in Critical Legal Studies (CLS) noting how CRT was in part a response to the perception that CLS analyses were too class-based and underestimated 'race', which for Critical Race Theorists is the major form of oppression in society. I then discuss some ethnic identity-specific varieties of CRT, before concluding with a brief consideration of its materialist and idealist forms.

The Voice of the Other

Postmodernism, transmodernism, and CRT all share a concern with the concept of the Voice of the Other. However, whereas postmodernism stresses multivocality, and transmodernism, the Voice of 'suffering Others', CRT, as we shall see, prioritizes the Voice of people of color.

Postmodernism

Elizabeth Atkinson (2002, p. 74) has defined postmodernism as:

- Resistance toward certainty and resolution;
- Rejection of fixed notions of reality, knowledge or method;
- Acceptance of complexity, of lack of clarity, and of multiplicity;
- Acknowledgement of subjectivity, contradiction and irony;
- Irreverence for traditions of philosophy or morality;
- Deliberate intent to unsettle assumptions and presuppositions;
- Refusal to accept boundaries or hierarchies in ways of thinking; and,
- Disruption of binaries which define things either/or.

Leading postmodernist Patti Lather (1991, p. 112) has argued that postmodernism 'celebrates multiple sites from which the word is spoken'.

Critiquing such celebration, Landon E. Beyer and Daniel P. Liston (1992) have posed the question, 'assuming that these voices will at least sometimes conflict', must we 'confront the status and validity of these multiple views—or simply assume they are all equally true (or false), equally revealing (or opaque)?' Moreover, as they point out, an emphasis on making sure that all voices are heard has made some postmodernists suspicious of 'community', since in communities, 'multiple voices' will be lost or silenced. Beyer and Liston conclude that, while we must value difference, multivocality can undermine action against oppression, including racism. As they put it:

> personal and social conditions need to be continually created, recreated, and reinforced that will encourage, respect, and value expressions of difference. Yet if the valorization of otherness precludes the search for some common good that can engender solidarity even while it recognizes and respects that difference, we will be left with a cacophony of voices that disallow political and social action that is morally compelling. If a concern for otherness precludes community in any form, how can political action be undertaken, aimed at establishing a common good that disarms partriarchy, racism and social class oppression? What difference can difference then make in the public space? (Beyer and Liston, 1992, pp. 380–381)

I would concur with Beyer and Liston that multivocality is a counterproductive concept in terms of collective struggle against racism and other forms of oppression (for a detailed Marxist critique of postmodernism, see Callinicos, 1989; Hill et al., 2002); see also Cole, 2008d, Chapter 5).

Transmodernism

As I understand them (Cole, 2008d, p. 69), transmodernism's defining features are:

- Rejection of totalizing synthesis;
- Critique of Modernity;
- Anti-Eurocentrism;
- Critique of Postmodernism;
- Analogic Reasoning: reasoning from 'OUTSIDE' the system of global domination[1];
- Reverence for (indigenous and ancient) traditions of religion, culture, philosophy and morality and Analectic Interaction[2]: not so much a way of thinking as a new way of living in relation to Others;
- Critique of (U.S.) Imperialism.

Transmodernist, David Geoffrey Smith (2003, p. 499) underlines the point that, unlike postmodernism, transmodernism actively seeks out not just Others, 'but…suffering Others'. For this reason, as I have argued elsewhere (Cole, 2008d, p. 75), it is theoretically and practically more progressive than postmodernism. Specifically, transmodernists are critical of Eurocentrism, the way in which the North is complicit in underdevelopment of the South, and the utter violence of that legacy which is not acknowledged formally within the West's dominant philosophical paradigms (Smith, 2003, p. 497) (for a critical appraisal of transmodernism from a Marxist perspective, see Cole, 2008d, Chapters 6 and 8; see also chapter 2 of this volume).

Critical Race Theorists, like transmodernists, have a total distrust of Eurocentrism, and focus on the situation of people of color in particular, rather than suffering Others per se, or the working class in general. People of color, it is argued, are *the* oppressed in the contemporary world (though, as we shall see later, this is limited to certain countries). CRT stresses in particular the importance of the *voice* of people of color, a voice framed by racism and at variance with the mainstream culture (for a discussion, see chapter 3 of this volume). In order to understand the origins of CRT, it is necessary to first look at Critical Legal Studies.

Critical Legal Studies

Critical Legal Studies (CLS) was a response to the law's role in protecting hierarchy and class. Thus for CLS, the oppressed are, in the classical Marxist sense, the working class.[3] According to Kimberlé Crenshaw et al. (1995a, p. 60), CLS was founded in 1976 at the Conference on Critical Legal Studies, comprised of a group of law teachers, students and some practitioners loosely organized as Left intellectuals and activists. CLT took its inspiration, Calvin Woodard (1986, p. 1) argues, from the teachings of the earlier 'legal realists' of the New Deal era.[4] In Crenshaw et al.'s (1995b, p. xviii) words, CLT aimed to expose and challenge 'the ways American law served to legitimize an oppressive social order'. Critical Legal Studies, as a movement, needs contextualizing. The United States (and the United Kingdom) had seen a successful attempt to consolidate capitalism with a significant lurch to the right in the 1970s. Crenshaw et al. (1995b, p. xviii) argue that '[b]y the late seventies, Critical Legal Studies existed in a swirl of formative energy, cultural insurgency, and organizing momentum', its conferences attracting hundreds of progressive law professors. CLS was not confined to the professors. In viewing law schools as work-places, and consequently as sites for organizing political resistance, (pp. xviii–xix):

> 'CLSers' actively recruited students and left-leaning law teachers from around the country to engage in the construction of left legal scholarship and law school transformation. (p. xix)

However, by the late 1970s, explicitly right-wing legal scholarship had also developed its own critique of conventional assumptions, making the law school 'an obvious site for ideological contestation as the apolitical pretensions of the "nonideological" center began to disintegrate' (p. xix). Perhaps the most unambiguous assertion within CLS of its adherence to the principle that law is an essentially *political* process appeared in a journal article written by Mark Tushnet (1981), secretary of the Conference on Critical Legal Studies from 1976 to 1985. After noting that a recent legal case confirmed 'the historical truth that the Supreme Court in its chambers does not sound very much different from Congress on the floor' (Tushnet, 1981, p. 413), and that for 'conceptual and political reasons, contemporary liberalism[5] cannot tolerate the existence of political institutions that mediate between the individual and the national government' (p. 421), and further that once 'a protected interest is identified, the Court must specify the procedures that due process requires' (p. 424), Tushnet states that when he reaches this point in the argument, he is invariably asked, 'Well, yes, but how would *you* decide the X case?' (ibid.). His answer is unequivocal: 'make an explicitly political judgment: which result is, in the circumstances now existing, likely to advance the cause of socialism?' (ibid.). Tushnet concludes that 'judges [should] not delude themselves into thinking that what they do has significance different from, and broader than, what every other political actor does' (pp. 425–426).

Credibility in the politics of CLS was enhanced as the disintegration of the center was confirmed by the election of Ronald Reagan in January 1981 (Margaret Thatcher had been elected in the United Kingdom in May 1979). Reagan's emphasis on 'supply-side' economics led to major cuts in federal funding which, though they were begun under the presidency of Jimmy Carter, accelerated drastically under Reagan. Coupled with the cuts in federal spending there were across the board tax cuts which favored the rich (the 'supply-side' claim being that benefits to the wealthy will 'trickle down' to benefit all). CLS was given a further boost with the appointment by Reagan of a number of Republican Federal judges.

In a seminal paper Duncan Kennedy (1982, p. 591), argued that law schools 'are intensely political places' that serve to reproduce the 'hierarchies of the corporate welfare state'. Kennedy's position was that in fact legal education *causes* hierarchy. Recalling Samuel Bowles and Herbert Gintis's (1976) correspondence principle, whereby that which goes in schools *corresponds* with structures and practices in capitalist industry, Kennedy put the case that law school treats students as if they are high school students at best and that legal education:

> supports [hierarchy] by analogy, provides it a general legitimating ideology by justifying the rules that underlie it, and provides it a particular ideology by mystifying legal reasoning. Legal education structures the pool of prospective lawyers so that their hierarchical organization seems inevitable and trains them in detail to look and think and act just like all the other lawyers in the system. (Kennedy, 1982, p. 607)

Kennedy was not suggesting, however, that law students are all pro-capitalist—on the contrary. As he points out, as well as service through law carried out with superb technical confidence, some students have a deep belief in law as an essentially progressive force, albeit much distorted by capitalism (p. 592).

There is, in addition, a contrasting more radical notion—the law as in essence superstructural—in the service of established interests (ibid.). Associated with the first belief is that Left-leaning students will be able to guarantee people their rights and bring about 'the triumph of human rights over mere property rights' (p. 598); associated with the second is the need for the student to reinterpret 'every judicial action as the expression of class interest' (p. 599), and to 'turn the tables exactly because she never lets herself be mystified by the rhetoric that is so important to other students' (p. 592). As Roberto Mangabeira Unger (cited in Long, 2005) puts it, recalling his days at law school:

> [w]hen we came, they [the law professors] were like a priesthood that had lost their faith and kept their jobs. They stood in tedious embarrassment before cold altars. But we turned away from those altars and found the mind's opportunity in the heart's revenge

Kennedy favors thinking about the law in a way 'that will allow students to enter into it, to criticize without utterly rejecting it, and to manipulate it without self-abandonment to an alien system of thinking and doing' (Kennedy, 1982, pp. 599–600). While he demonstrates a sound awareness of issues of gender and 'race' and of his own position in 'the system of class, sex, and race (as an upper middle-class, white male)' (p. 609), Kennedy's focus, as a CLS scholar, is decidedly on social class. Aware of his rank in the professional hierarchy—a Harvard professor, Kennedy advocates a strategy to build 'a left bourgeois intelligentsia that might one day join together with a mass movement for the radical transformation of American society' (p. 610). 'Without such a cadre of bourgeois intellectuals', he goes on, given the success of the dominant ideology, and the fact that most Americans define themselves as middle class (p. 610), 'it is unlikely that a mass movement could ever be permanently successful in the United States'. Kennedy's form of anti-capitalist struggle is inclusive, and as further evidence of this, he argues that study groups, the 'core of a law-school organizing strategy' (p. 611) should make an effort to find 'radical feminist networks' and 'activist groups in the black community', as well as 'local labor-union insurgents' (p. 613). Kennedy also favors hiring women, minority and working class candidates, until these groups are reasonably represented in the faculty (p. 615). However, for Kennedy, '[t]o understand law is to understand [the struggle for power through law] as an aspect of class struggle and as an aspect of the human struggle to grasp the conditions of social justice' (p. 599). The Conference on Critical Legal Studies, Kennedy points out, organizes annual meetings aimed at Marxist and non-Marxist law teachers who are attempting to radicalize their work lives (Kennedy, 1982, p. 612). This might include equalizing all salaries in the school, including secretaries and janitors, and everyone spending one month per year performing a job in a different part of the hierarchy (p. 615).

Kennedy aims to help the students organize against the authoritarian classroom; for curriculum change to reduce right-wing political bias and its incapacity for alternative forms of practice; and to establish politically sensitive legal services clinics for poor people, operated by the law school (p. 612).[6]

CRT: The Beginnings

As Crenshaw et al. (1995b, p. xix) explain, CRT emerged in the interstices of the political and institutional dynamic created by the disintegration of the center ground, and represented an attempt to inhabit the space between two very different ideological and intellectual formations. First, CRT was a reaction to Critical Legal Studies, criticizing CLS for its emphasis on class and economic structure, and 'its failure to come to terms with the particularity of race' (p. xxvi). As Cornel West (1995, p. xi) puts it, 'critical legal studies writers… "deconstructed" liberalism, yet seldom addressed the role of deep-seated racism in American life'. Second, CRT 'sought to stage a simultaneous encounter with the exhausted vision of reformist civil rights scholarship' (Crenshaw et al., 1995b, p. xix), its liberal tradition having directly impinged on the lived experiences of people of color in the law schools (ibid.). As Crenshaw et al. (ibid.) put it:

> We both saw and suffered the concrete consequences that followed from liberal legal thinkers' failure to address the constrictive role that racial ideology plays in the composition and culture of American institutions, including the American law school

As they go on, CLS 'provided a language and a practice for viewing the institutions in which we studied and worked both as sites of and targets for our developing critique of law, racism, and social power' (ibid.). Critical Race Theorists sought 'a left intervention into race discourse and a race intervention into left discourse' (ibid.).[7]

CRT is generally thought to derive from a number of sources: the work of Frantz Fanon who perceived the world to be divided between 'two different species', a 'governing race' and 'zoological' natives (Fanon, 1963, pp. 40–42, cited in Mills, 1997, p. 112); and from the writings of W. E. B. Du Bois, for whom 'the problem of the Twentieth Century [was] the problem of the color-line' (Du Bois, 1903).[8] Richard Delgado and Jean Stefancic (2001, p. 4) also add as influential the names of [the humanist Marxist] Antonio Gramsci; of [post-structuralist] Jacques Derrida; of [abolitionist and women's rights activist] Sojourner Truth; [abolitionist and equality advocate] Frederick Douglass; [labor leader and civil rights activist] Cesar Chavez; and [civil rights activist and civil rights activist and campaigner for workers' rights] Martin Luther King, Jr. (see the Conclusion of this volume for a discussion of King), as well as the Black Power and Chicano movements of the sixties and early seventies; and radical feminism.

While there is no definitive birth date to the movement, according to Kimberlé W. Crenshaw (cited in Lawrence III et al., 1993, p. 4) CRT's social origins lay in a student boycott and alternative course organized at the Harvard Law School in 1981. The main objective of the protest was to persuade the administration to increase the number of tenured professors of color in the faculty. Derrick Bell, Harvard's first African-American professor, had left to take up the position of Dean of the law school at the University of Oregon (ibid.). This left only two professors of color in the Harvard Law School. Students demanded that Harvard begin to rectify the situation by hiring a person of color to teach the course, 'Race, Racism and American Law' that had regularly been taught by Bell (Lawrence III et al., 1993, p. 4). When it became clear that the administration was not prepared to meet their demand, the students organized an alternative course, with leading academics and practitioners invited on a weekly basis to lecture and lead discussion on Bell's book (Bell, 1973) which had the same title as the course (ibid.), and which had regularly been taught by Bell (Isaksen, 2000, p. 696). Crenshaw et al. (1995b, p. xxi) describe this alternative course as 'in many ways the first institutionalized expression of Critical Race Theory'. The course led to study groups, conferences and a proliferation of CRT scholarship (Isaksen, 2000, p. 696). As Crenshaw et al. (1995b, p. xxii) put it '[t]he Alternative Course reflected—as well as helped to create—the sense that it was meaningful to build an oppositional community of left scholars of color within the mainstream legal academy'.[9]

A turning point in the CLS/CRT distinction occurred at the 1986 CLS Conference. Feminist legal theorists who organized this conference invited scholars of color to facilitate discussions about 'race'. The latter posed the question, 'what is about the whiteness of CLS that discourages participation by people of color?' which produced a pitched and heated debate, a dialogue to which some CLS scholars were resistant (xxiii).

Another defining moment occurred during the 1987 CLS conference, entitled 'The Sounds of Silence', during discussions about what Crenshaw et al. (p. xxiv.) refer to as 'racialism'. As used here this does not refer to the erstwhile everyday meaning of the term,[10] but to 'theoretical accounts of racial power that explain legal and political decisions which are adverse to people of color as mere reflections of underlying white interest' (ibid.). For CLS scholars this was too deterministic an account and resonated with their distrust of 'vulgar' Marxism, a form of Marxism where the economic base *determines* the superstructure (of which law is a constituent element) (ibid.).[11] As Crenshaw et al. (ibid.) explain, during the 1980s, CLS scholars had shown skepticism toward this form of Marxism. They argued instead that, rather than being a mere instrument of class interests, 'the law is an active instance of the very power politics it purports to avoid and stand above' (Crenshaw et al., 1995b, xxiv). Drawing on these premises, Critical Race Theorists began to think of their project as 'uncovering how law was a constituent element of race itself: in other words, how law *constructed* race' (p. xxv). Crenshaw et al. (ibid.) explain how the dialectical model of CLS informed CRT:

> we accepted the crit [CLS] emphasis on how law produces and is the product of social power and we cross-cut this theme with an effort to understand this dynamic in the context of race and racism

This was something, Crenshaw et al. (ibid.) argue, CLS was for the most part unable to do.

The 1987 CLS Conference plenary saw large numbers of scholars of color articulate how 'institutional practices and intellectual paradigms functioned to silence insurgent voices of people of color' (p. xxvi). Another scholar of color responded to this point by presenting the view that scholars of color needed to stop complaining and start to write (ibid.). This conference was attended by scholars such as Richard Delgado, Mari Matsuda and Pat Williams who were to become leading lights in the movement and led to a symposium issue of the Harvard Law Review that included a lot of the early signature CRT pieces (Delgado, 2008, personal correspondence). In addition the Boalt Hall Coalition for a Diversified Faculty held a series of lectures in the late 1980s and, like the 1987 conference, led to tangible consequences including some published papers and a national strike for diversity (ibid.) This highly successful nationwide law student strike on April 6, 1989 involved at least thirty schools (Kidder, 2003, p. 37), and entailed students boycotting class and carrying out teach-ins to protest against 'discrimination based on race, gender, economic class, and sexual orientation within America's law schools' (Herrera, 2002, pp. 80–81, cited in Kidder, 2003, p. 37).

Crenshaw et al. (1995b, p. xxvi) conclude that the '1986 and 1987 CLS conferences...marked significant points of alignment and departure, and should be considered the final step in the preliminary development of CRT as a distinctly progressive critique of legal discourse on race'. CRT is *aligned* with CLS in the sense of 'radical left opposition to mainstream legal discourse' (pp. xxvi–xxvii), and *differentiated* in its focus on 'race' (p. xxvii).

The key formative event, however, according to Crenshaw et al. (1995b, p. xxvii) was the founding of the CRT workshop in 1989. The workshop was financed by a grant provided by David Trubek, a founding member of the Critical Legal Studies Conference and drew together thirty five law scholars. The organizers, principally Kimberlé Crenshaw, Neil Gotanda, and Stephanie Philips, coined the term 'Critical Race Theory' to make it clear that their work was located in the 'intersection of critical theory and race, racism and the law'. This workshop featured a couple days of discussion aimed at finding out whether all the participants shared a common core of principles and ideas, and if so which ones. The ideas in Crenshaw's influential paper entitled 'Race, Reform and Retrenchment: Transformation and Legitimization in Antidiscrimination Law', which had been published in the *Harvard Law Review* in 1988 were central to these discussions.

By the mid-1990s, according to leading Critical Race Theorist, Gloria Ladson-Billings (2005, p. 57) legal scholars had written more than 300 leading law review articles and a dozen books on CRT.

Cementing the pivotal relationship between CRT and law, a pioneering (1995) article by Ladson-Billings and William Tate traces the development of what they refer to as the *Property Issue* in the United States. This, they note to be a defining feature of the society encompassing a sequence of exploitative events which began with the removal of native Americans from the land, continued through military conquest of the Mexicans to the construction of Africans as property.

Ladson-Billings and Tate begin by stating that racism is endemic and deeply ingrained in American life and that the cause of African-American underachievement is 'institutional and structural racism' (Ladson-Billings and Tate, 1995, p. 55). They then go on argue that segregation continues in U.S. schools with African-American and Latino/a students still faring badly as compared to white students. Following Delgado et al. (1989), they argue that the dominant group justifies its power with stories that 'construct reality in ways to maintain their privilege' (Ladson-Billings and Tate, 1995, p. 58). Assuming that Ladson-Billings and Tate meant the dominant economic group, rather than white people per se, Marxists would be in full agreement with these observations. Moreover, 'justifying power with stories' has a certain resonance with structural Marxist, Louis Althusser's (1971) concept of the interpellation of subjects.[12]

CRT: Identity-Specific Varieties

Since the analytical tools of CRT come from a range of epistemologies there is no linear progression overall in the theory. Some CRT scholars (e.g., Dixson, 2006) have likened CRT to jazz which can be in unison without being in unison, and which grew out of African-Americans' resistance to oppression and their struggle for equality. CRT, however, encompasses far more than the experiences of African-Americans. Rebecca Tsosie (2005–2006, p. 22) has argued that 'CRT is a jurisprudence of possibility precisely because it rejects standard liberal frameworks *and precisely because it seeks to be inclusive of different groups and different experiences*' (my emphasis). In this section, I will try to give a flavor of some identity-specific varieties of CRT, and will look at some unique features of each, including some tensions, as well as overall commonalities.

LatCrit and Black Exceptionalism

In 1997, the first LatCrit Conference took place. At that conference, Juan Perea challenged the black/white binary (Chang and Gotanda, 2007, p. 1014). In a subsequent paper, Perea (1997, pp. 231–237) advocated five LatCrit axioms:

- Our goal must be the most broad understanding of equality. It must be a full equality, admitting of no qualifications or impediments;
- The concepts of 'Race' and 'Racism' must be amplified to promote Latina/o equality;
- The concept of civil rights is so dominated by the black/white binary understanding of American racial identity that it is currently of little utility for Latinos;

- 'National Origin' is not a helpful concept in understanding discrimination against Latinos/as nor in redressing such discrimination;
- The concepts of ethnicity and ethnic identity may be the most appropriate set of group traits for amplifying our understanding of race in a way that discrimination against Latinos/as can be recognized and understood.

Perea's analysis needs to be seen as a response to 'black exceptionalism'. As Delgado and Stefancic (2001, p. 69) point out, '[e]xceptionalism holds that a group's history is so distinctive that placing it at the center of analysis is, in fact, warranted'. While exceptionalism could apply to any group, Delgado (2007, personal correspondence) suggests that most of the CRT movement is 'caught up in the traditional black-white binary', a stance which he totally opposes. 'Black exceptionalism' has been described by Angela Harris (Espinoza and Harris, 2000, p. 440), on the other hand, as 'an important truth', since it is both 'an intellectual and a political challenge to LatCrit theory' (p. 444). As she notes:

> As an intellectual claim, black exceptionalism answers Perea's criticism of the black-white paradigm by responding that the paradigm, though wrongly making 'other non-whites' invisible, rightly places black people at the center of any analysis of American culture or American white supremacy.... In its strongest form, black exceptionalism argues that...Indians, Asian Americans, and Latino/as do exist. But their roles are subsidiary to, rather than undermining, the fundamental binary national drama. As a political claim, black exceptionalism exposes the deep mistrust and tension among American ethnic groups racialized as 'nonwhite'. (ibid.)

Robert Chang and Neil Gotanda (2007, p. 1017) state that 'there has been surprisingly little engagement with the challenge that the Black exceptionalism claim poses to LatCrit', and that when it has been taken up beyond a footnote reference, the authors have tended to be black (ibid.). As they conclude, although this might be done for reasons of 'politeness' (ibid.), this 'can...weaken in the long run coalition with other groups' (ibid.). I would agree with Delgado and Stefancic (2001, p. 70), who have also engaged with 'black exceptionalism, that '[t]he black-white—or any other—binary paradigm of race...weakens solidarity, reduces opportunities for coalition, deprives the group of the benefits of the others' experiences' (p. 70). The black-white paradigm, Delgado and Stefancic argue, 'simplifies analysis dangerously' (ibid.). As they explain, in the context of the United States:

> Few blacks will be yelled at and accused of being foreigners...Few will be told that if they don't like it here, they should go home. Few will be ridiculed on account of their unpronounceable last names or singsong accent. Few will have a vigilante, police officer, teacher, or social worker demand to see their papers, passport, or green card. By the same token, few Asian-looking people will be accused of being welfare leeches or having too many children out of wedlock. (Delgado and Stefancic, 2001, pp. 69–70)

This scenario is, of course, very U.S.-specific, and applicable to African-Americans and Asian-Americans (in the U.S. sense of the term). In different contexts, black people *are* accused of being foreign and told to 'go home', *are* ridiculed because of their names or accents (as are Asians and others in various contexts). Black people *are* suspected of being 'illegal immigrants', Asian-looking people *are* accused of being reliant on welfare and having too many children.

This differential racialization, however, is recognized by Delgado and Stefancic when they note that racialization varies historically, as 'each disfavored group...has been racialized in its own individual way and according to the needs of the majority group at particular times in its history' (Delgado and Stefancic, 2001, pp. 69). From a Marxist perspective, the racialization of specific groups does not occur according to the needs of 'the majority group', but according to the needs of the minority group of (white) capitalists and their politician supporters. For Marxists, the racialization (and xeno-racialization) of specific groups needs to be related to the mode of production (this is discussed in the next chapter of this volume).

Chang and Gotanda (2007, p. 1029) conclude that 'it is crucial that we examine closely the positions and actions taken in past and present racial coalitions' and that part of a LatCrit agenda should be 'the political imperative to re-examine 'Black exceptionalism' (ibid.). While I would agree with these sentiments which, indeed, accord with my preferred wide-ranging definition of racism outlined in chapter 2 of this volume, I would be totally against, as I am sure would all Critical Race Theorists, any marginalization of the very brutal racism directed historically and contemporaneously at African-Americans, from the days of slavery onward.

Asian-American Jurisprudence

The Asian-American experiences of racism are different in a number of ways from those of the African-American and Latina/o communities. Some of these relate to racist stereotyping, which can impact negatively on other minority ethnic groups. For example, at the first annual Asian Pacific American Law Professors Conference that took place in 1994, Sumi Cho referred to the phenomenon of 'racial mascotting' (Chang and Gotanda, 2007, p. 1020): 'using Asian Americans as a person of color as a mascot for an argument that really isn't about Asian Americans at all' (BAMN, 2001). Frank Wu, cited in BAMN (2001) explains:

> with the affirmative action debate as it's typically carried out, Asian Americans are brought into this debate as a wedge group. Instead of bringing us in to expand the dialogue, instead of bringing us in to recognize that we are American citizens, that we are minorities, that we have a stake in this process, and that civil rights laws protected us, what often happens is Asian Americans are brought in to this debate and held up, and that message is heard over and over again that they made it, why can't you.

In addition to the scapegoating of others, another problem with this 'model minority' is identified by Wu (cited in ibid.): it is a false stereotype and we should be suspicious of any stereotype no matter how positive because of what it can conceal (this issue is taken up in my definition of racism in chapter 2 of this volume). For example, the myth causes a backlash for Asian-Americans. Describing Asian-Americans as hard working very quickly becomes 'unfair competition'. Similarly, designating Asian-Americans as good at math and science leads to notions that they are 'nerdy and geeky, and can't be lawyers, they can't be managers, they lack of people skills' (cited in ibid.). Finally Wu notes that when Asian-Americans are praised for having strong families and strong family values—nuclear families that stay together, they are then criticized for being 'too clannish, too ethnic, too insular, not mixing enough, self-segregating' (cited in ibid.) (the dangers of 'seemingly positive attributes' are discussed in the context of my definition of racism in chapter 2 of this volume). Robert Chang (2000, p. 359) notes another negative response which results from the 'model minority' myth: 'when we try to make our problems known, our complaints of discrimination or calls for remedial action are seen as unwarranted and inappropriate'. As Chang (p. 360) concludes, 'the danger of the model minority myth [is that it] renders the oppression of Asian Americans invisible'.

Wu (cited in ibid.) goes on to make reference to ways in which the racism directed at Asian-Americans is similar to that which other racialized groups experience. For example, there is a glass ceiling, whereby those Asians who have the same qualifications and the same educational background as their white peers, and who work in the same types of jobs earn less money. There is also housing segregation and hate crime directed against Asian-Americans. Robert Chang (2000, p. 354), a teacher of legal writing, at the time he wrote the chapter, gives some personal experiences: his valid Ohio driver's license not being good enough to let him return to his own country, the United States, after a brief trip to Canada; being refused service at a service station, when driving in the South (ibid.); being stopped in New Jersey for suspicion of being in possession of a stolen vehicle (pp. 354–345)—first two police officers, then four, perhaps 'afraid that I might know marital arts', and not receiving an apology, when his license and registration papers had been checked out (p. 355).

Wu also emphasizes differential experiences *within* Asian-American communities. Chinese Americans and Japanese Americans, he argues, have incomes that cluster toward the top, while South East Asians and Filipinas/os are clustered toward the bottom, their socio-economic status being more similar to that of African-Americans than white Americans (cited in BAMN (2001)). Chang and Gotanda (2007, p. 1027) also note what they describe as 'the fault lines with regard to the coherence of the Asian-American racial category' with respect to Filipinas/os and 'the differential racialization of South Asians following 9/11) (Islamophobia is discussed in chapter 2 of this volume). Finally, Chang (2000, p. 360) notes the disproportionate rates of poverty, when compared to national rates, of Laotians, Hmong, Cambodians, and Vietnamese.

Native Jurisprudence

Native American CRT scholars have been active in addressing indigenous people's rights, sovereignty and land claims. Here we also witness sameness and disparity. Christine Zuni Cruz (2005) elucidates her feelings that CRT is a lived experience for indigenous peoples:

> I realize, as people of color, as people who love our families, our people, we don't engage in critical race analysis in the abstract. We are a part of the analysis. We experience the impact of race, color, and culture in the context of power. It is in the stories of our fathers, our sisters, our children, all those within our communities. It is a compelling story that must be told and must be included in the analyses of the law that our profession engages in. If we don't speak them, no one else will.

Castagno and Lee (2007, p. 4) note that Native Jurisprudence or 'TribalCrit includes tenets and principles that 'are culturally specific to Indigenous people and communities'. The term 'TribalCrit' (Tribal Critical Race Theory) was coined by Bryan Brayboy (2005); Brayboy (p. 429–430) lists its nine tenets follows:

1. Colonization is endemic to society.
2. U.S. policies toward Indigenous peoples are rooted in imperialism, white supremacy, and a desire for material gain.
3. Indigenous peoples occupy a liminal space that accounts for both the political and racialized natures of our identities.
4. Indigenous peoples have a desire to obtain and forge tribal sovereignty, tribal autonomy, self-determination, and self-identification.
5. The concepts of culture, knowledge, and power take on new meaning when examined through an Indigenous lens.
6. Governmental policies and educational policies toward Indigenous peoples are intimately linked around the problematic goal of assimilation.
7. Tribal philosophies, beliefs, customs, traditions, and visions for the future are central to understanding the lived realities of Indigenous peoples, but they also illustrate the differences and adaptability among individuals and groups.
8. Stories are not separate from theory; they make up theory and are, therefore, real and legitimate sources of data and ways of being.
9. Theory and practice are connected in deep and explicit ways such that scholars must work toward social change.

As Castagno and Lee (2007, p. 5) put it following Brayboy, 2005 and Tsosie, 2000, '[t]ribal nations have a unique history and a unique political relationship with the federal government, and both of these factors must be central to analyses'.

Tsosie takes the specific example of Native Hawaiians. The Hawaiian monarchy was overthrown by a group of American insurgents, with the help of the U.S. military, in 1893 (Tsosie, 2005–2006, p. 32). Tsosie (p. 43) argues that '[e]xamining the concept of justice through the CRT lens of "equality" demands that we consider the notion of "equality of nations"' as a possibility to achieve justice and reconciliation (ibid.). For Native Hawaiian people:

> are not 'Indians'. They are not merely 'indigenous peoples'. They are the Kanaka Maoli and they are the contemporary descendants of the Kingdom of Hawaii, which never voluntarily surrendered to the United States. (p. 42)

Referring to U.S. policy toward Indigenous peoples in general within Federal Indian Law, Tsosie explains that Indian tribes had only a 'right to occupancy' rather than the territorial rights of a 'real' nation (p. 43). Moreover, because tribes were perceived to be in a rudimentary state of governance, they were deemed to be 'wards' of the federal government (ibid.). 'The notion that Indian nations', she continues, are not fully socialized or civilized also creeps into these cases, masked as a fear that tribes will not respect the "civil rights" of non-Indians, or even non-member Indians' (p. 44). This was not always the case Tsosie points out, and cites treaties that were signed in a spirit of diplomacy between Europeans and Native *Nations* in the eighteenth century (pp. 44–45).

Because the privilege and power resides in federal and state governments rather than with tribal governments, she concludes, the ensuing problems such as lack of meaningful tribal jurisdiction over assaults and domestic violence between non-Indians and Indians, for example, are treated as a 'social condition' (p. 44), 'amenable perhaps for federal funding to "study" the problem' (ibid.). 'But no one seriously questions the validity of the basic structure' (ibid.).

Citing Cheryl Harris (1993) Tsosie (p. 43) goes on to make the case that 'formal equality' can and often does disadvantage people of color, and that CRT transcends formal justice and builds 'on the truth of the political, social, economic and spiritual conditions experienced by a people, to analyze alternative possibilities to achieve justice' (Tsosie, 2005–2006, p. 43). Tsosie's solution is for Native people 'to define, assert, protect, and insist upon respect for the right to be what they always have been: distinctive governments and societies, autonomous and free' (p. 45). This means resisting the colonial enterprise, which uses external power to define the 'other' as subordinate to the colonial nation, and taking back power and constructing tribal sovereignty from *within* Native societies (ibid.) (the restoration of indigenous rights in the Bolivarian Republic of Venezuela, including collective ownership of land, and the right to an intercultural and bilingual education is discussed in chapter 7 of this volume).

These distinctive differences in the Native American experiences, any more than those of the other racialized groups in the United States, do not, of course, detract from the commonalities of racism experienced by racialized communities there and, indeed, worldwide. Overall, similarities and differences between African-Americans, Latina/o-Americans, Asian-Americans and Native Americans underline Dixson's (2006) aforementioned analogy with jazz—in unison and out of unison at the same time. As will be argued throughout this volume, however, these differences and similarities, to be fully understood, need to be related to and articulated with ongoing changes in the capitalist mode of production.

CRT has a number of other variants which are not specifically 'ethnic' variants, and therefore, fall outside the remit of this book. These include Critical Race Feminism (Harris, 2001, p. xx); and 'a...queer-crit interest group'; (Delgado and Stefancic, 2001, p. 6) (see Delgado and Stefancic, 2000 for discussions of these variants).

Materialist and Idealist CRT

CRT, according to Richard Delgado (2001), has an 'idealist wing' and a 'materialist wing'. The former, he argues, is concerned with discourse analysis, and maintains that 'racism and discrimination are matters of thinking, attitude, categorization, and discourse' (ibid.). In focusing on 'words, symbols, stereotypes and categories' (Delgado, 2003, p. 123), combating racism means that we have to 'rid ourselves of the texts, narratives, ideas and meaning that give rise to it and convey' negative messages about specific groups (ibid.). Like post-structuralism the analytic tools are discourse analysis (Delgado, 2003, p. 123). (For an analysis of post-structuralism, see Cole, 2008d, Chapter 4.)

Materialist CRT, on the other hand, focuses on material factors and views racism as a 'means by which society allocates privilege, status and wealth' (Delgado, 2001, p. 2). Materialist CRT scholars are interested in factors 'such as profits and the labor market' (Delgado, 2003, p.124). Such scholars are also interested in international relations and competition and in 'the interests of elite groups, and the changing demands of the labor market' and how this benefits or disadvantages 'racial' groups historically (ibid.). The legal system is key in sanctioning or punishing racism, depending on its larger agenda (ibid.). Materialist CRT, then, has an affinity with both Max Weber and Weberian analysis of capitalism, and with Marxism. Delgado, who is one of the few exponents of materialist CRT (this will become clear throughout this volume—see, for example, the section entitled 'Delgado and Going Back to Class' in chapter 2), argues that CRT is 'almost entirely dominated' by the idealist wing of CRT and that this means that there are 'huge deficiencies' in our understanding of institutional racism and ways in which the law is being used to serve dominant groups (pp. 124–125). From a Marxist perspective, while in focusing in part on capitalism and profits and the labor market, the 'materialist wing' is clearly preferable to the 'idealist wing', Delgado's 'huge deficiencies' can only be fully addressed, I will argue throughout this book, by a full engagement with Marxist thinking.

In this chapter, I began by briefly noting the relationship between post-modernism, transmodernism and CRT, before tracing CRT's origins in

Critical Legal Studies. Both CLS and CRT, I have shown, had very firm origins in the law. I then discussed CRT's identity-specific varieties, and concluded with a brief discussion of its materialist and idealist wings. In the next chapter, I will critique, from a Marxist perspective, CRT's two major tenets (to which a number of other tenets relate): the first is that the concept of 'white supremacy' better describes oppression based on 'race' in contemporary societies than does the concept of 'racism'; the second is the belief in 'race' rather than social class as the primary contradiction in society.

Chapter 2

White Supremacy and Racism; Social Class and Racialization

In this chapter[1] I critique from a Marxist perspective two of CRT's central tenets, namely the favoring of the concept of 'white supremacy' over racism, and the prioritizing of 'race' over class as the primary form of oppression in society. Next I offer my own wide-ranging definitions of racism and the Marxist concepts of racialization and xeno-racialization, arguing that these formulations are better suited in general to understanding and combating racism in the contemporary world, than is the CRT concept of 'white supremacy'.

Tenet I: 'White Supremacy' Rather than 'Racism'[2]

In 1989, bell hooks noted:

> As I write, I try to remember when the word racism ceased to be the term which best expressed for me the exploitation of black people and other people of color in this society and when I began to understand that the most useful term was white supremacy. (hooks, 1989, p. 112)

Eight years later, Charles Mills (Mills, 1997), following on from the European social contract theorists of the seventeenth and eighteenth centuries (Hobbes, Locke, Hume, Rousseau), and more contemporaneously and more specifically Carole Pateman's (1988) *Sexual Contract*,[3] set out to explain how and why people of color in the modern era are treated unequally in relation to white people. The book opens with the following statement:

> White supremacy is the unnamed political system that has made the modern world what it is today. (Mills, 1997, p. 1)

In chapter 1 of this volume, I looked at Chang and Gotanda's (2007) analysis of LatCritTheory and Asian Jurisprudence. It should be pointed out here that one of their primary concerns about differentiation between minority communities was that 'White supremacy often gets lost' (Chang and Gotanda, 2007, p. 1017), since for them, '[i]n the American context, one must never lose sight of White supremacy' (p. 1019; see also pp. 1020, 1021, 1023, 1027).

For Mills, 'white supremacy' is 'the basic political system that has shaped the world for the past several hundred years' (1997, p. 1) and 'the most important political system of recent global history' (ibid.). The Racial Contract, according to Mills (1997, p. 33) 'designates Europeans as the privileged race'. To underline the point that Mills sees 'white supremacy' as a political system in its own right, and that the Racial Contract is both 'real' and 'global' (p. 20), Mills asserts:

> Global white supremacy...is *itself* a political system, a particular power structure of formal or informal rule, socioeconomic privilege, and norms for the differential distribution of material wealth and opportunities, benefits and burdens, rights and duties. (p. 3)

However, according to Mills (p. 32), it is the economic dimension of the Racial Contract which is the *most* salient, since it is 'calculatedly aimed at economic exploitation'. In command of the Racial Contract is the 'white-supremacist state' (p. 82).

Arguing in the same vein as Mills (1997), influential UK Critical Race Theorist, David Gillborn (2005) reconceptualizes 'white supremacy' not in terms of 'the usual narrow focus on extreme and explicitly racist organizations' (pp. 485–486), but in terms of 'a central and extensive form of racism that evades the simplistic definitions of liberal discourse' (p. 492), one that 'is normalized and taken for granted' (p. 486). For Gillborn (2008, p. 35) white supremacy relates to 'the operation of forces that saturate the everyday mundane actions and policies that shape the world in the interests of White people'. 'White supremacy' signifies 'a deeply rooted exercise of power that remains untouched by moves to address the more obvious forms of overt discrimination' (Gillborn, 2005, p. 492) to which the concept of 'racism' usually refers. In other words, Gillborn believes that 'white supremacy' should replace the concept of 'racism' because the concept of 'racism' tends to put the focus on overtly racist practices that 'are by no means the whole story'. The concept of 'racism' thus 'risks obscuring a far more comprehensive and subtle form of race politics' (p. 491)—that which he believes is captured by his articulation of 'white supremacy'. As he explains in relation to education, 'white supremacy' involves 'the routine assumptions that structure the system' and 'encode a deep privileging of white students and, in particular, the legitimization, defence and extension of Black inequity' (p. 496).

Namita Chakrabarty and John Preston, also prominent in UK CRT, (2006a, p. 1) have ascribed 'white supremacy', along with capitalism itself, the status of an 'objective inhuman [system] of exploitation and oppression'. In a similar vein, Gillborn (2006a, p. 320) has commended Ansley's (1997, p. 592) definition of 'white supremacy' as 'a political, economic and cultural *system* in which whites overwhelmingly control power and material resources' (my emphasis).

I would argue that there are four significant problems with the term 'white supremacy'. The first is that it can direct critical attention away from modes of production; the second that it homogenizes all white people together as being in positions of power and privilege; the third is that it inadequately explains what I have referred to elsewhere (Cole, 2007b, p. 13) as 'non-colour-coded racism'; and the fourth that it is totally counter-productive as a political unifier and rallying point against racism. I will deal with each in turn.

Directing Attention Away from Modes of Production

While, for Marxists, it is certainly the case that there has been a *continuity* of racism for hundreds of years, the concept of 'white supremacy' does not in itself *explain* this continuity, since it does not need to connect to modes of production and developments in capitalism. It is true that Mills (1997) provides a wide-ranging discussion of the history of economic exploitation, and that Preston (2007) argues that CRT needs to be considered alongside Marxism. However, unlike Marxism, there is no *a priori* need in CRT for-mulations to connect with capitalist modes of production. In Marxist par-lance, the mode of production refers to the *combination* of forces (human labor power and the means of production) and the relations of production (primarily the relationship between the social classes). This combination means that the way people relate to the physical world and the way people relate to each other are bound together in historically specific, structural and necessary ways. As Marx (1859) put it:

> The totality of these relations of production constitutes the economic struc-ture of society, the real foundation, on which arises a legal and political super-structure and to which correspond definite forms of social consciousness. The mode of production of material life conditions the general process of social, political and intellectual life.

Critical Race Theorists do not analyze these crucial relationships. Thus Gillborn (e.g., 2005, 2006a) is able to make the case for CRT and 'white supremacy' without providing a discussion of the relationship of racism to capitalism. The Marxist concept of racialization, however, *does* articulate with modes of production. Examples of the ways in which it does this are discussed later in this chapter.

The Homogenization of All White People

While for Critical Race Theorists 'white supremacy' primarily describes the *structural* dimension of 'white power', 'white privilege' mainly refers to the *day-to-day practices* that arise directly or indirectly from 'white supremacy'. However, both interact with each other (Delgado, personal correspondence, 2008), and both have structural and day-to-day practical implications.

Thus immigration restrictions would be part of the structural dimension of the 'white supremacist' state (ibid.), but with obvious day-to-day practical manifestations. From a Marxist perspective, it is, of course, the poor and dispossessed rather than the rich and powerful, whose entry into other (richer) countries is restricted (although this exclusion is dependent on capitalists' relative need for cheap labor).

Delgado (ibid.) gives an example of the *practical* nature of 'white privilege' when 'store clerks put change directly in the upraised palms of white customers but lay the coins down on the counter for blacks or Latinos/ Latinas'. For Critical Race Theorists, such practices are also enshrined *structurally* in 'white supremacist' societies. For Marxists, the class element is crucial. Rich people of color are less likely to get their change thrust on the counter. Moreover, well-off people of color will tend to shop in more 'upmarket' stores, and will be more disposed to the use of plastic as a form of payment.

Critical Race Theorists believe that *all* white people are beneficiaries of 'white supremacy' and 'white privilege'. Gillborn (2008, p. 34) states that while they are not all active in identical ways, and do not all draw similar advantages, '*[a]ll White-identified people are implicated in*...[relations of shared power and dominance]—...*they* do *all benefit, whether they like it or not*'.

Sabina E. Vaught and Angelina E. Castagno (2008, p. 99) would appear to hold similar views and refer to 'the ways in which power over others...benefits Whites individually and collectively' (p. 99), and specifically emphasize white privilege's 'structural nature' (p. 100). They argue (2008, p. 96) that 'Whiteness as property is a concept that reflects the conflation of Whiteness with the exclusive rights to freedom, to the enjoyment of certain privileges, and to the ability to draw advantage from these rights'. Following Cheryl Harris (1993, p. 1721) they state that 'to be identified as white' was 'to have the property of being white. Whiteness was the characteristic, the attribute, the property of free human beings'. 'In this way', Vaught and Castagno (2008, p. 96) continue, 'individual White persons came to exercise, benefit from, and mutually create and recreate a larger structural system of collective, institutional White privilege' (ibid.). Again, following Harris (1993, p. 1762), they refer to 'the continued right to determine meaning' (Vaught and Castagno, 2008, p. 101), and make reference to Peggy McIntosh's (1988) notion of systemic 'arbitrarily-awarded' privilege (Vaught and Castagno, 2008, p. 99). They conclude that the societal systems 'that sustain the reign of White race privilege are peopled and the concurrent, interactive acts of individuals and systems inexorably reinforce and entrench pervasive racial power across institutions, sites and events' (p. 96). 'White racial power', they claim, 'permeates every institution' (p. 101).

When Gillborn makes reference to McIntosh's 'famously listed 50 privileges' (Gillborn, 2008, p. 35), and describes them as 'privileges that accrue from being identified as White', he has seriously misunderstood McIntosh's list. In merely describing the privileges as accruing from being identified as white, he decontextualizes and dehistoricizes her analysis. In actual fact, McIntosh contextualizes white privilege with respect to her social class position as a white academic with respect to her 'Afro-American co-workers, friends, and acquaintances' with whom she comes into 'daily or frequent contact in this particular time, place, and line of work' (p. 293).[4] Homogenizing the social relations of all white people ignores, of course, this crucial social class dimension of privilege and power.

Mills (1997, p. 37) acknowledges that not 'all whites are better off than all nonwhites, but [argues that]...as a statistical generalization, the objective life chances of whites are significantly better'. While this is, of course, true, we should not lose sight of the life chances of millions of working class white people. To take poverty as one example, in the United States, while it is the case that the number of black people living below the poverty line is some three times that of whites, this still leaves over 16 million 'white but not Hispanic' people living in poverty in the United States (U.S. Census Bureau, 2007). This is indicative of a society predicated on racialized *capitalism*, rather than indicative of a white supremacist society. While the United States is witnessing the effects of the New Racial Domain (Marable, 2004—see below) with massively disproportionate effects on black people and other people of color, white people are also affected. In the United Kingdom, there are similar indicators of a society underpinned by rampant racism, with black people currently twice as poor as whites, and those of Pakistani and Bangladeshi origin over three times as poor as whites (Platt, 2007).[5] Once again, however, this still leaves some 12 million poor white people in the United Kingdom, who are, like their American counterparts, on the receiving end of global neoliberal capitalism. The devastating effects of social class exploitation and oppression are masked by CRT blanket assertions of 'white supremacy' and 'white privilege'.

There are further problems with the homogenization of all whites. First it masks essential power relations in capitalist societies. For Marxists, the ruling class are *by definition* those with power since it is they who own the means of production, and the working class, in having to sell their labor power in order to survive, are (also by definition) the class largely without power. The manifestations of this major power imbalance in the capital/labor relation massively affects relative degrees of privilege in capitalist, the aforementioned rates of poverty being just one.

Lack of power for the working class is particularly evident in countries like the United States and the United Kingdom where that class has been successfully interpellated (Althusser's concept of interpellation, outlined in chapter 1 of this volume). Moreover, some of the very privileges that poor white people possess are in a very real sense *compensatory* privileges. For example, Delgado (2008, personal correspondence) has introduced the concept of 'paltry privileges' to describe those 'privileges' that whites enjoy that compensate for the fact that they are living in impoverished conditions with low paid jobs, unpaid bills and poor life chances. Alpesh Maisuria and I (Cole and Maisuria, 2008) made a similar point when referring to the success of soccer in keeping white workers in line:

> Ruling class success in maintaining hegemony in the light of the disparity of wealth and the imperial quest was displayed in England during the 2006 World Cup by the number of St. George flags signifying a solid patriotism in run-down (white) working class estates, on white vans, on dated cars exhibiting a 'proud to be British' display. In addition, as economically active migrant workers from Eastern Europe enter the UK (a great benefit for capital, and for the middle strata who want their homes cleaned or renovated cheaply), the (white) working class, who spontaneously resist neo-liberalism by resisting working for low wages that will increase their immiseration, need to be assured that they 'still count'. Hence the ruse of capital is to open the markets, and the role of sections of the tabloid media is to racialize migrant workers to keep the (white) working class happy with their lot with the mindset that 'at least we are not Polish or Asian or black, and we've got our flag and, despite everything, our brave boys in Iraq did us proud.

In Althusser's words, their response is: 'That's obvious! That's right! That's true!' (Althusser, 1971, p. 173). In this case the homogenization of all whites obfuscates the *ideological* element of the capital/labor relation. While it is undoubtedly true that racism and xeno-racism (see below) have penetrated large sections of the white working class, resulting in racist practices that contribute to the hegemony of whites, and while it is clearly the case that members of the (predominantly though not exclusively) white ruling class are the beneficiaries of this, it is certainly not white people as a whole who hold such power (Cole and Maisuria, 2008). For example, sections of the white working class in England have voted for the fascist British National Party (BNP) at recent elections *precisely* because they feel that they are treated with less equality than others (Cruddas et al., 2005).

There are thus a number of problems with homogenizing all white people. Attempts to do this ignore capitalist social relations, which are infused with the crucial dimensions of social class, power and ideology.

Non-Color-Coded Racism

Mills acknowledges that there were/are what he refers to as '"borderline Europeans"—"the Irish, Slavs, Mediterraneans, and above all, of course, Jews"' (Mills, 1997, pp. 78–79), and, elsewhere (Mills, 2007, p. 249) that the Irish may have been 'the first systematically racialized group in history'. He also notes there also existed '*intra*-European varieties of "racism"' (p. 79; see also Perea et al.). However, he argues that, while there remain 'some recognition of such distinctions 'in popular culture'—he gives examples of an '"Italian" waitress' in the TV series *Cheers,* calling a WASP character

'Whitey' and a discussion in a 1992 movie about whether Italians are really white (p. 79)—he relegates such distinctions primarily to history. While he is prepared to 'fuzzify' racial categories (p. 79) with respect to 'shifting criteria prescribed by the evolving Racial Contract' (p. 81) and to acknowledge the existence of 'off-white' people at certain historical periods (p. 80), he maintains that his categorization—'white/nonwhite, person/subperson' 'seems to me to map the essential features of the racial polity accurately, to carve the social reality at its ontological joints' (p. 78), whereby white = person; non-white = non-person.

It is my view that this does not address current reality. The exclusive forefronting of people of color militates against an understanding of non-color-coded-racism.

Marxist 'race' theorist Robert Miles (1987, p. 75) argues that racialization is not limited to skin color:

> The characteristics signified vary historically and, although they have usually been visible somatic features, other non-visible (alleged and real) biological features have also been signified.

I would like to make a couple of amendments to Miles' position. First, I would want to add 'and cultural' after, 'biological'. Second, the common dictionary definition of 'somatic' is 'pertaining to the body', and, given the fact that people can be racialized on grounds of symbols (e.g., the hijab), I would also want this to be recognized in any discussion of social collectivities and the construction of racialization. Miles' Marxist analysis of racism is discussed at length later in this chapter.

Racism directed at white people is not new and has a long history. To take the case of Britain, for example, there has been a long history of non-color-coded racism directed at the Irish (e.g., Mac an Ghaill, 2000)[6], at the Gypsy Roma Traveler (GRT) communities (e.g., Puxon, 2005)—the fastest growing minority ethnic constituency in Europe (Doughty, 2008), and increasingly at the Muslim communities or those perceived to be Muslim.

Anti-Gypsy Roma Traveler Racism

As the UK Schools Minister Lord Adonis points out many GRT pupils are among the lowest-achieving in UK schools (see also Department for Children, Schools and Families, 2008), and the situation is not improving. Fear of racism and bullying means that many children and families are too afraid to identify themselves (*The Guardian*, March 11, 2008). In a 2007 survey of over 200 GRT children by the Children's Society, 63% said they had been physically attacked and 86% report that they have been verbally abused as a result of their ethnic origin. The report pointed out that poor attendance and achievement in schools may reflect these high levels of racism (R. Davis, 2008, p. 5). Although many GRT children now go to primary school, very few enter secondary education. Of those surveyed, one-third had dropped out by the time they reached 10, and three-quarters by the time they reached 13 (ibid.) Rowenna Davis (ibid.) concludes that the problems are likely to grow as GRT populations grow with EU expansion.

The United Kingdom's most popular (right-wing) tabloid—read predominantly by the working class (and here I am using working class in its sociological sense—see chapter 1, Note 3 of this volume), *The Sun* both captures and creates working class racism (see Cole, 2004b, pp. 161–163). In its March 24, 2008 edition, the front page headlines screamed, 'Gypsy Hell for Tessa'. The article was referring to the fact that sixty-four travelers had 'set up camp just yards from the country home of government minister Tessa Jowell' (p. 1). Above the headline was the caption, 'Easter Holiday Invasion'. Other descriptors included '30 caravans swarmed on to the...field' (p. 1); 'Gypsy Nightmare'; 'crafty gypsies' (p. 4); 'these families' (p. 8). The travelers had actually bought the site—'secretly' according to *The Sun* (p. 1), and needed to move in while council enforcement officers were on holiday (p. 1). The land is zoned for agricultural use, and has no planning permission for homes (p. 4). As one of the travelers told *The Sun* 'we...have been evicted thousands of times while on the road...over the past seven years. We just want a permanent home' (p. 4). The travelers may be able to seek redress from the UK Human Rights Act 1998 which entitles everyone to 'the right to own property' and 'the right to respect for private and family life'. *The Sun* described the Act whose main planks are the right to life; freedom from torture and degraded treatment; freedom from slavery and forced labor; the right to liberty; the right to a fair trial; the right not to be punished for something that wasn't a crime when you did it; the right to respect for private and family life; freedom of thought, conscience and religion; freedom of expression; freedom of assembly and association; the right to marry or form a civil partnership and start a family; the right not to be discriminated against in respect of these rights and freedoms; the right to own property; the right to an education; and the right to participate in free elections (The Human Rights Act, 1998) as 'detestable' (p. 8).

Islamophobia

Islamophobia is a major form of racism in the modern world. It is important to stress that Islamophobia, like anti-GRT racism, is not necessarily triggered by skin color—it is often sparked off by one or more (perceived) symbols of the Muslim faith. It has, of course, intensified since the 9/11 attack on the Twin Towers in New York in 2001, and the suicide bombings of July 7, 2005 (7/7) in Britain (when a coordinated attack was made on London's public transport system during the morning rush hour). The invasion of Iraq has, of course, further intensified Islamophobia (see chapter 6 of this volume; see also Cole and Maisuria, 2007, 2008). People who appear to be of the Islamic faith are immediately identified as potential terrorists and in Britain are five times more likely to be stopped and searched than a white person (Dodd, 2005). In many ways, the racialization of the Muslim communities of Britain, which involves pathologizing and scapegoating is similar to the way in which 'black youth' were racialized in the 1970s and 1980s (e.g., Hall et al., 1978; see also Cole, 1986b, pp. 128–133). The role of the media is important here too. Peter Oborne recently presented a UK TV program (based on Oborne and Jones, 2008; see Mason, 2008). The program commissioned a survey of nearly 1,000 articles written since 2000, noting the content and context of articles pertaining to Muslims and Islam. This showed that 69 percent of the articles presented Muslims as a source of problems not just in terms of terrorism but also with respect to cultural issues. It found that 26 percent of the articles portrayed Islam as danger-ous, backward or irrational (Mason, 2008). As Mason (2008) concludes the shortcoming of the pamphlet (Oborne and Jones, 2008) 'is that it fails to link it to other aspects of government policy: namely the whipping up of fear of terrorist attacks and using the "war against terror" to justify the wars in Iraq and Afghanistan as well as numerous attacks on democratic rights'.

Xeno-racism

In addition to the above forms of racism which are not necessarily color-coded, Britain, for example, is also witnessing, in the current period, a new form of racism, a racism which has all the hallmarks of traditional racism, but which impacts on recently arrived groups of people, a non-color coded racism which has been described by Sivanandan (2001, p. 2) as xeno-racism. He defines it as follows:

> a racism that is not just directed at those with darker skins, from the former colonial countries, but at the newer categories of the displaced and dispos-sessed whites, who are beating at western Europe's doors, the Europe that displaced them in the first place. It is racism in substance but xeno in form—a racism that is meted out to impoverished strangers even if they are white. It is xeno-racism.

Sivanandan (ibid.) has underlined xeno-racism's *economic* basis:

> under global capitalism which, in its ruthless pursuit of markets and its sancti-fication of wealth, has served to unleash ethnic wars, balkanise countries and displace their peoples, the racist tradition of demonisation and exclusion has become a tool in the hands of the state to keep out the refugees and asylum seekers so displaced—even if they are white—on the grounds that they are scroungers and aliens come to prey on the wealth of the West and confound its national identities. The rhetoric of demonisation, in other words, is racist, but the politics of exclusion is economic.

But not all are excluded. Indeed, in the United Kingdom in the twenty-first century, the enlarged European Union provides an abundance of cheap Eastern European labor, and there is substantial evidence of xeno-racism in Europe (Fekete, 2009; see also Cole, 2004b, 2007b; Cole and Maisuria,

2007, 2008 for a discussion of xeno-racism in the United Kingdom).[7] This includes escalating verbal and physical assault. The BBC News Web Site notes resentment among locals (2008a) and reports six instances of racist attacks on the Polish communities this year alone (at the time of writing it is only half-way through the year) (BBC News, 2008b, 2008c, 2008d, 2008e, 2008f, 2008g; see also BBC News, 2005a, 2005b; *Belfast Today*, 2006; BBC News, 2007).[8] Eastern European workers are also pathologized by the media, and certain tabloids (in particular *The Sun*) have unleashed anti-Polish and other anti-Eastern European rhetoric on a regular basis.

While CRT certainly reminds us that 'race' is central in sustaining the current world order, and that we must listen to the voices of people oppressed on grounds of racism, it does not and cannot make the necessary connections to understand and challenge all forms of racism. The limits to the CRT argument are that it restricts racism ('white supremacy' in CRT terms) to a set of practices directly related to skin color. In claiming that 'with the exception of Nazi Germany...borderline Europeans...were not subpersons in the full technical sense and would all have been ranked ontologically above genuine nonwhites' (Mills, 1997, p. 80), Mills is seriously underestimating the actual and potential virulence of present day non-color coded racism. In focusing on issues of color and being divorced from matters related to capitalist requirements with respect to the labor market, CRT is ill-equipped to analyze the discourses non-color-coded racism. Marxism, on the other hand, in having its focus on the capitalist economy, *does* provide explanations.

White Supremacy as a Unifier and Political Rallying Point

Here is the platform of *Race Traitor* (2005), an organization that takes the dangers of 'white supremacy' to their limits and that calls for the abolition of whiteness:

What We Believe

The white race is a historically constructed social formation. It consists of all those who partake of the privileges of the white skin in this society. Its most wretched members share a status higher, in certain respects, than that of the most exalted persons excluded from it, in return for which they give their support to a system that degrades them.

The key to solving the social problems of our age is to abolish the white race, which means no more and no less than abolishing the privileges of the white skin. Until that task is accomplished, even partial reform will prove elusive, because white influence permeates every issue, domestic and foreign, in US society.

The existence of the white race depends on the willingness of those assigned to it to place their racial interests above class, gender, or any other interests they hold. The defection of enough of its members to make it unreliable as a predictor of behavior will lead to its collapse.

RACE TRAITOR aims to serve as an intellectual center for those seeking to abolish the white race. It will encourage dissent from the conformity that maintains it and popularize examples of defection from its ranks, analyze the forces that hold it together and those that promise to tear it apart. Part of its task will be to promote debate among abolitionists. When possible, it will support practical measures, guided by the principle, Treason to whiteness is loyalty to humanity.

I have argued elsewhere (Cole, 2008d, p. 115) that the *style* in which the organization, *Race Traitor*'s ideological position is written is worryingly reminiscent of Nazi propaganda, and seriously open to misinterpretation: that it could be interpreted as meaning the abolition of white *people*. In fact, it is made clear above and in the book of the same name (Ignatiev and Garvey, 1996) that this is not the case.[9] However, when one taps in 'Race Traitor' on a *Google* search, it is the above statement *written* by the organization 'White Traitor', which comes up first. I am not questioning the sincerity of the protagonists of 'the abolition of whiteness', nor suggesting in any way that they are anti-white *people*—merely questioning its extreme vulnerability to misunderstanding.[10]

Antiracists have made some progress, in the United Kingdom at least, after years of 'establishment' opposition, in making *antiracism* a mainstream rallying point, and this is reflected, in part, in legislation (e.g., the (2000) Race Relations Amendment Act).[11] Even if it were a good idea, the chances of making 'the abolition of whiteness' a successful political unifier and rallying point against racism are virtually non-existent. For John Preston (2007, p. 13), '[t]he abolition of whiteness is...not just an optional extra in terms of defeating capitalism (nor something which will be necessarily abolished post-capitalism) but fundamental to the Marxist educational project as praxis'. Indeed, for Preston (2007, p. 196) '[t]he abolition of capitalism and whiteness seem to be fundamentally connected in the current historical circumstances of Western capitalist development'. From a Marxist perspective, coupling the 'abolition of whiteness' to the 'abolition of capitalism' is a worrying development which, if it gained ground in Marxist theory in any substantial way would most certainly undermine the Marxist project, even more than it has been undermined already (for an analysis of the success of the Ruling Class in forging consensus to capitalism in the United Kingdom, see Cole, 2008g, 2008h). Implications of bringing the 'abolition of whiteness' into schools are discussed in chapter 7 of this volume. As is argued in this volume, racism, xeno-racism, racialization, and xeno-racialization, when informed by Marxism, are far more conducive to understanding racism in contemporary societies than is the CRT concept of 'white supremacy'. 'White supremacy', I believe, should be restricted to its conventional usage.

Tenet II: 'Race' Not Class as the Primary Contradiction

As noted in chapter 1 of this volume, the key formative event in the establishment of CRT was the CRT workshop in 1989 which made clear CRT's location in critical theory, and crucially 'race', racism and the law.

Ladson-Billings and Tate (1995, p. 62) underline the CRT belief in 'race' as primary by aligning their scholarship and activism with the philosophy of Marcus Garvey, who believed that any program of emancipation would need to be built around the question of 'race' first. As they observe, Garvey is clear and unequivocal:

> In a world of wolves one should go armed, and one of the most powerful defensive weapons within reach of Negroes is the practice of race first in all parts of the world. (cited in ibid.)

Similarly, Mills (2003, p. 156) rejects both what he refers to as the 'original white radical orthodoxy (Marxist)' for arguing that social class is the primary contradiction in capitalist society, and the 'present white radical orthodoxy (post-Marxist/postmodernist)' for its rejection of any primary contradiction. Instead, for Mills (ibid.) 'there is a primary contradiction, and...it's race'. For Crenshaw et al. (1995b, p. xxvi) 'subsuming race under class' is 'the typical Marxist error'.

Mills (2003, p. 157) states that '[r]ace [is] the central identity around which people close ranks' and that there is 'no transracial class bloc'. Given the way in which neoliberal global capitalism unites capitalists throughout the world on lines that are not necessarily color-coded, this statement seems quite extraordinary.

'Race', Mills goes on, is 'the stable reference point for identifying the 'them' and 'us' which override all other 'thems' and 'us's' (identities are multiple, but some are more central than others)', (Mills, 2003, p.157) while for Crenshaw et al. (1995b, p. xxvi), although they acknowledge that 'race' is socially constructed, something with which Marxists would fully concur,[12] 'race' is 'real' since 'there is a material dimension and weight to being "raced" in American society'. 'Race', Mills concludes is 'what ties the system together, and blocks progressive change' (Mills, 2003, p.157). For Marxists, it is capitalism that does this.

Mills (1997, p. 111) argues that '[w]hite Marxism [is] predicated on colorless classes in struggle', and suggests that 'European models of radicalism, predicated on a system where race is much less domestically/internally important (race as the external relation to the colonial world), operate with a basically raceless (at least nominally) conceptual apparatus'(Mills, 2003, pp. 157–158). 'Race', he states, 'then has to be "added on"' (p. 158). Claiming that Marxism is 'largely seen as dead' (Mills, forthcoming, 2009),[13] Mills states that he would like to think that 'a modified historical materialism that takes race seriously instead of seeing it as merely epiphenomenal to class' can explain 'white supremacy' (ibid.). If so, such a Marxism, he concludes, 'does have to be a theoretically revised one' (ibid.), not 'the class-reductionist Marxism' that he designates as '"white Marxism", a Marxism that fails to recognize the import and social reality of race' (ibid.).

My response to Mills' desire for Marxism to explain 'white supremacy', for which in the closing paragraph of Mills, 2003 (p. 247) Mills states that he has 'left open the door' but is unsure if it can (Mills, forthcoming, 2009), is that I do not believe there will be a Marxist explanation of 'white supremacy', since, as outlined in the previous section of this chapter, the concept is incompatible with Marxism. With respect to Mills' call for a non 'class-reductionist Marxism', and his proclaimed sympathy with the idea that Marxism 'ultimately provides the most promising theoretical tool for understanding the genesis and persistence of racism' (forthcoming, 2009), I would answer in the following way. Mills use of the adverb 'ultimately' and his statement that 'this seems more of a project in progress that a successfully completed one' (Mills, forthcoming, 2009) does not do justice to long-standing and wide ranging US-based (e.g., Torres and Ngin, 1995; Zarembka, 2002; Darder and Torres, 2004; Marable, 2004; Scatamburlo-D'Annibale and McLaren, 2004, 2008, 2009), and UK-based Marxist analyses of 'race' and racialization (e.g., Miles, 1982, 1984, 1987, 1989, 1993; Sivanandan, 1982, 1990; Callinicos, 1993; Cole, 2004a, 2004d, 2006a, 2006b, 2008c, 2008d, 2009a, 2009d; Cole and Virdee, 2006; Virdee, 2009a, 2009b). (The Marxist concepts of racialiazation and xeno-racialization are discussed in more depth later in this chapter; see also chapter 4 of this volume for a discussion of Marxism and antiracism, and chapter 5 for a discussion of Marx and 'race'.)

Mills (2003, p. 158) invites readers to

Imagine you're a white male Marxist in the happy prefeminist, pre-postmodernist world of a quarter-century ago. You read Marcuse, Miliband, Poulantzas, Althusser. You believe in a theory of group domination involving something like the following: The United States is a *class* society in which class, defined by *relationship to the means of production*, is the *fundamental* division, the bourgeoisie being the *ruling* class, the workers being *exploited* and *alienated*, with the state and the juridical system *not* being neutral but part of a superstructure to maintain the existing order, while the *dominant ideology* naturalizes, and renders invisible and unobjectionable, class domination.

This all seems a pretty accurate description of the United States in the twenty-first century, but for Mills (ibid.) it is 'a set of highly controversial propositions'. He justifies this assertion by stating that all of the above 'would be disputed by mainstream political philosophy (liberalism), political science (pluralism), economics (neoclassical marginal utility theory), and sociology (Parsonian structural-functionalism and its heirs)' (ibid.). While this is true, my response to this would be, well, of course it would be disputed by mainstream philosophers, pluralist political scientists, neoclassical economists and functionalist sociologists, all of which, unlike Marxists, are, at one level or another, apologists for capitalism.

The Salience of Social Class

Social class, I would argue, albeit massively racialized (and gendered) is the system upon which the maintenance of capitalism depends (Kelsh and

Hill, 2006, Hill, 2007b, Hill, 2008b, 2009c). It is possible, though extremely difficult because of the multiple benefits accruing to capital of racializing workers (not least forcing down labor costs), and the unpaid and underpaid labor of women as a whole, to imagine a capitalist world of 'racial' (and gender) equality. It is not logically possible for capitalism to exhibit social class equality (see Kelsh and Hill, 2006; see also Hill, 2007b; Hill et al., 2008; Kelsh et al., 2009). Without the extraction of surplus value from the labor of workers, capitalism cannot exist (Marx, 1887; see the appendix to chapter 8 of this volume). There are four caveats I need to add to this fore-fronting of social class.

First, I fully agree with Critical Race Theorists (e.g., Gotanda, 1995; Delgado and Stefancic, 2001, pp. 21–23) that we should reject 'color blindness',[14] the belief that 'one should treat all persons equally, without regard to their race' (Delgado and Stefancic, 2001, p. 144). As Delgado and Stefancic explain:

> Critical race theorists... hold that color blindness will allow us to redress only extremely egregious racial harms, ones that everyone would notice and condemn. But if racism is embedded in our thought processes and social structures as deeply as many crits believe, then the 'ordinary business' of society—the routines, practices, and institutions that we rely on to effect the world's work—will keep minorities in subordinate positions. Only aggressive, color-conscious efforts to change the way things are will do much to ameliorate misery (Delgado and Stefancic, 2001, p. 22).

Second, I agree with David Roediger (2006, p. 3) that Left commentators are wrong to announce the end of Du Bois's 'century of the color line' (e.g., Gilroy, 2000; Patterson, 2000, cited in Roediger, 2006, p. 4). Paul Gilroy (e.g., 2004) has more recently expressed a somewhat over-optimistic in my view belief in multicultural 'conviviality'. As I will argue in chapter 3 of this volume, one of CRT's strengths is its insistence of the all-pervasive existence of racism in the world.

Third, while I totally reject the views of those contemporary Left theorists (e.g., Apple, 2005, 2006) which promote the idea that 'race' and class are *equivalent* (for a Marxist critique, see, for example, Kelsh and Hill, 2006), I would insist that arguments made that, because of the centrality of class, the Left should not concern itself with issues of racism are fundamentally flawed. Thus, I believe that Adolph Reed's arguments that '[a]s a political strategy, exposing racism is wrongheaded, and at best an utter waste of time', and that '[racism] is the political equivalent of an appendix' to social class (Reed, 2005a; see also Reed, 2005b) are extremely dangerous and not conducive to progressive struggle.

Fourth, my critique of CRT accords with that of Darder and Torres (2004) (misleadingly lumped together with Reed by Roediger, 2006, p. 4) in two major respects. The first is that, as I indicated earlier in this chapter, it is my view that 'race' is a social construct and has no scientific validity. The second, as I will argue later in this chapter, is that the Marxist concept of racialization provides a more convincing explanation of racism than CRT notions of 'white supremacy', and is necessary in order to understand the multiple manifestations of racism *and* their relationship to modes of production. In the context of these multiple manifestations, the debate between class *or* racism becomes redundant, in that for Marxists the struggle is against racialized (and gendered) capitalism.

Delgado and Going Back to Class

When CRT was originally envisioned, it was to be an intersection of 'racial theory' and activism against racism. However, a number of CRT theorists today are frustrated at the turn CRT has made from activism to academic discourse, and this has led to a reappraisal of the significance of social class.

As we saw in chapter 1 of this volume, Delgado (2003) has put forward a materialist critique of the discourse-focused trend of recent CRT writings which focus more on text and symbol and less on the economic determinants of Latino/a and black racial fortunes. Delgado's paper was the subject of a symposium in 2005, run by *The Michigan Journal of Race and Law*, entitled, *Going Back to Class: The Re-emergence of Class in Critical Race Theory*. Somewhat surprisingly, given Mills' published comments on Marxism above (see also Pateman and Mills, 2007), Mills said he favored the combination of Marxism and CRT, which forms a kind of 'racial capitalism.'[15] He said he agreed with Delgado on the belief, central to CRT, that class structure keeps racial hierarchy intact. The working class is divided by 'race', Mills said, to the advantage of the upper class, which is mainly composed of white elites (Hare, 2006), a position very familiar to Marxist analysts.[16]

At the same symposium, Angela Harris said CRT is essential in exposing how interconnected class, 'race' and sex can be: 'We need to pay attention to the intersections and understand how complicated these issues are,' (cited in Hare 2006). As an example, she referenced the affirmative action disputes in higher education. The often-cited argument that working-class whites are being rejected in favor of middle-class blacks and Latinos—who, the argument goes have a better chance of acceptance regardless of 'race'—is looking at class based solely on income (cited in ibid.). 'What CRT exposes is that class also needs to be looked at in terms of access to wealth and the racialization of class' (cited in ibid.).

As for the future of CRT, Delgado envisions a new movement of CRT theorists to recombine discourse and political activism. 'I'm worried that the younger crop of CRT theorists are enamored by the easy arm-chair task of writing about race the word and not race in the world', Delgado concluded. 'A new movement is needed'. For Marxists, these are promising developments and point towards a possible alignment between CRT and Marxism. However, any future alignment would need to have at its core a structural analysis of capitalism and capitalist social relations, combined with a critique thereof (I return to this in the Conclusion to this volume). I now turn to a consideration of racism and Marxism before relating racism to the Marxist concept of racialization.

Racism and Marxism

It should be clear from the above analysis that I favor a wide-ranging definition of racism and racialization in order to account for changes in racism which accompany changes in the capitalist mode of production. Shortly, I would like to offer my preferred definition of racism. I should point out at this stage that this definition is different to that favored by a number of other Marxists who prefer the analysis of Robert Miles. Miles and his associates are totally against inflation of the concept of racism. Miles (1989) argues *against* inflating the concept of racism to include actions and processes as well as discourses. Indeed, he argues that 'racism' should be used to refer exclusively to an *ideological* phenomenon, and not to exclusionary practices. He gives three reasons for this. First, exclusionary practice can result from both intentional and unintentional actions (Miles, 1989, p. 78). I would argue, however, that the fact that racist discourse is unintentional does not detract from its capacity to embody racism. For its recipients, effect is more important than intention (see my definition of racism later in this chapter). Second, such practices do not presuppose the nature of the determination, for example, the disadvantaged position of black people is not necessarily the result of racism (ibid.). However, the fact that the 'disadvantaged position of black people is not necessarily the result of racism' is addressed by Miles' own theoretical approach, a class-based analysis which also recognizes other bases of unequal treatment. Therefore, I would argue, this recognition does not need the singling out that Miles affords it. Miles' third reason for making racism exclusively ideological is that there is a dialectical relationship between exclusion and inclusion: to exclude is simultaneously to include and vice versa, for example the overrepresentation of African-Caribbean children in 'special schools' for the 'educationally subnormal' (ESN) in the 1960s involves both exclusion from 'normal schools' and inclusion in ESN schools (ibid.). I do not see the purpose of this attempt to privilege inclusion. The simultaneous inclusion of black people entailed by exclusion is, by and large, a negative inclusion, as in the case of Miles' own example of ESN schools. There are, of course, situations where exclusion on account of the application of positive labels leads to positive consequences for those thus labeled. The way monarchies and aristocracies are perceived is an obvious example. They are excluded from everyday life but included in very elite settings with multiple positive benefits. I fail to see how Miles' observation about the dialectical relationship between exclusion and inclusion informs an analysis of racism.

While I understand Miles' desire to retain a Marxist analysis, and not to reify racism (since describing actions and processes as 'racist' may forestall an analysis of various practices in different historical periods of capitalist development), it is my view, as I attempted to demonstrate in this book, that it is *precisely* the Marxist concept of racialization (and xeno-racialization) that enables, and indeed *requires*, a persistent and constant analysis of the multiple manifestations of racism in different phases of the capitalist mode of production in different historical periods. Indeed, I try to show in this chapter that, contrary to Miles, not only should racism be inflated to incorporate actions, processes and practices, but that it should, in fact, be inflated considerably to include a *wide range* of actions, processes and practices. Miles' position on not inflating the concept of racism retains a fervent following in the Department of Sociology at the University of Glasgow where Miles first expounded his views on racism and racialization. I attended a workshop there in 2006, entitled *What can Marxism teach Critical Race Theory about Racism* (Centre for Research on Racism, Ethnicity and Nationalism (CRREN) Department of Sociology, University of Glasgow). Some Marxist sociologists who attended were quite insistent on defending Miles' position, and stressed the need to use *Marxist* terminology rather than the concept of racism (though no such terminology was generally forthcoming).[17] One contributor went as far as to express the view that 'there is not a lot of racism out there'. Another, also following Miles, stated that racism should be narrowed down, and confined to the level of *ideas*, and that *actions* should not be described as racist. The same delegate found the concept of racialization problematic, adding that people 'magically becoming racialized' is meaningless. Another delegate argued that, whereas once people were sure what racism was; now both in the United Kingdom and globally, it is difficult to understand what racism is. Miles and the Marxist defenders of his position are right to be wary of any tendency to call everything 'racist' and thereby to foreclose discussion. However, in my view, there are grounds for believing that if an action or process is perceived to be racist then it probably is. Indeed this is enshrined in the excellent UK Race Relations (Amendment) Act (2000). What I think should distinguish a Marxist analysis of racism is the attempt to relate various instances of racism and (xeno-) racialization to different stages in capitalist development (some examples are given later in this chapter), but also to relate them to political and other ideological factors. This is not to say that all individual or institutional instances of racism and racialization are reducible to the economy (Miles acknowledges this as a functionalist position), but that racism and racialization in capitalist countries needs to be understood in terms of stages in capitalist development. I take the position that there are striking similarities in actions and processes of racism and (xeno-) racialization directed against different people in differing economic, political and ideological circumstances. This is *not* to claim that racism is primary and that all else flows from it, which is the position of Critical Race Theorists and was the fear of one delegate at the workshop, but to stress the need for retaining the concept of racism, widening it and relating it to developments in capitalism.

Racism Defined

Contemporary racism, both in its ideological forms and material practices, might best be thought of as a matrix of biological and cultural racism. I would argue that, in that matrix, racism can be based on genetics (as in notions of white people having higher IQs than black people: see Herrnstein &

Murray, 1994[18]; and more recently Frank Ellis (Gair, 2006)[19] or on culture (as in contemporary manifestations of Islamophobia). Sometimes, however, it is not easily identifiable as either (e.g., 'Britain for the British'), or is a combination of both. A good example of the latter is when Margaret Thatcher, at the time of the Falklands/Malvinas war, referred to the people of that island as 'an island race' whose 'way of life is British' (Short and Carrington, 1996, p. 66). Here we have a conflation of notions of 'an island race' (like the British 'race' who, Mrs. Thatcher believes, built an empire and ruled a quarter of the world through its sterling qualities; (Thatcher, 1982, cited in Miles, 1993, p. 75)) and, in addition, a 'race', which is culturally, like 'us': 'their way of life is British'.

There are also forms of racism which are quite unintentional, which demonstrates that you do not have to be *a* racist (i.e., have allegiance to far-Right ideologies) to be racist, or to be implicated in generating racism consequences. Thus when somebody starts a sentence with the phrase 'I'm not racist but...', the undertone means that the next utterance will always be racist. The use by some people in the United Kingdom, *out of ignorance,* of the term 'Pakistani' to refer to everyone whose mode of dress or accent, for example, signifies that they might be of Asian origin is another example of unintentional racism. The use of the nomenclature 'Paki', on the other hand, I would suggest, is generally used in an intentionally racist way because of the generally known negative connotations attached to the word in the United Kingdom.

Racism, as practices, can also be overt, as in racist name-calling in schools, or it can be covert, as in racist mutterings in school corridors.

For Miles (1989, p. 79), racism relates to social collectivities identified as 'races' being 'attributed with negatively evaluated characteristics and/or represented as inducing negative consequences for any other'. Here I would also want to inflate Miles' definition to include, following Smina Akhtar, 'seemingly positive attributes'. However, ascribing such attributes to an 'ethnic group' will probably ultimately have racist implications, for example the subtext of describing a particular group as having a strong culture might be that 'they are swamping *our* culture'. This form of racism is often directed at people of South Asian origin who are assumed to have close-knit families and to be hard working, and therefore in a position to 'take over' *our* neighborhoods.[20]

In addition, attributing something seemingly positive—'*they* are good at sport'—might have implications that 'they are not good' at other things. People of African-Caribbean origin, in racist discourse, are thought to have 'no culture' or a *different* culture, and to thus also pose a threat to 'us'. In education this is something that facilitates the underachievement of working class African-Caribbean boys who are thought to be (by some teachers) less academically able, and 'problems'. Stereotypes and stratifications of ethnic groups are invariably problematic and, at least potentially, racist.

Racism can be dominative (in the form of direct and oppressive state policy) as in the apartheid era in South Africa or slavery in the United States, or it can be aversive, where people are segregated, excluded or cold-shouldered on grounds of racism (Kovel, 1988), or where they are routinely treated less favorably in day-to-day interactions. In certain situations, racism may well become apparent given specific stimuli. For example, racism in the media, as in the above examples of anti-gypsy/Roma/Traveller racism and Islamophobia, can actually generate racism. Similarly, racist sentiments from a number of peers who might be collectively present at a given moment can facilitate racist responses. Racist chants and other racist reactions at soccer matches are an obvious example.

Elsewhere (Cole, 2008d, pp. 117–119), I have advocated a definition of racism which includes cultural as well as biological racism, intentional as well as unintentional racism; 'seemingly positive' attributes with probably ultimately racist implications as well as obvious negative racism; dominative racism (direct and oppressive) as opposed to aversive racism (exclusion and cold-shouldering) (cf. Kovel, 1988) and overt as well as covert racism. Finally, and crucially in the context of this chapter, racism can be non-color-coded. All of these forms of racism can be individual or personal, and they can be brought on, given certain stimuli. These various forms of racism can also take institutional forms (see chapter 5 of this volume for a discussion of institutional racism), and there can, of course, be permutations among them. I would argue, therefore, that, in order to encompass the multifaceted nature of contemporary racism, it is important to adopt a broad concept and definition of racism, rather than a narrow one, based, as it was in the days of the British Empire, or pre–civil rights United States, for example, on notions of overt biological inferiority, even though there may also have been implications of cultural inferiority. I believe that the above conception and definition of racism both theoretically and practically better depicts racism in contemporary Britain (and elsewhere) than CRT notions of 'white supremacy'. From a Marxist perspective, in order to understand and combat racism, however, we must relate it to historical, economic and political factors. I now turn to the Marxist concept of racialization which makes the connection between racism and capitalist modes of production and thus is able to relate to these factors, namely the real material contexts of struggle. While the extraction of surplus value defines the whole capitalist process (see the appendix to chapter 8 of this volume), permutations of these factors vary greatly, and it is for this reason that the definition of racism has to be wide-ranging.

Racialization

Miles (1987) has defined racialization as an ideological[21] process that accompanies the appropriation of labor power (the capacity to labor), where people are categorized falsely into the scientifically defunct notion of distinct 'races'. As Miles puts it, the processes are not explained by the fact of capitalist development (a functionalist position). Racialization, like racism, is socially constructed. In Miles' (1989, p. 75) words, racialization refers to 'those instances where social relations between people have been structured by the signification of human biological characteristics [elsewhere in the same book, Miles (1989, p. 79) has added cultural characteristics] in such a way as to define and *construct* differentiated social collectivities' (my emphasis). Consistent with my own definition of racism I would want to add, in addition to 'the biological' and 'the cultural', the other dimensions outlined above. '[T]he process of racialization', Miles states, 'cannot be adequately understood without a conception of, and explanation for the complex interplay of different modes of production and, in particular, of the social relations necessarily established in the course of material production' (Miles, 1987, p. 7). It is this articulation with modes of production which makes the concepts of racialization and xeno-racialization inherently Marxist.[22]

Racialization and the British Empire

I have developed the links between modes of production and racialization in the United Kingdom at length elsewhere (e.g., Cole, 2004a, 2004b). For example, in Cole, 2004a, p. 39, noting that racialization is historically and geographically specific, I argued that, in the British colonial era, implicit in the rhetoric of imperialism was a racialized concept of 'nation'. British capitalism had to be to regenerated in the context of competition from other countries, and amid fears that sparsely settled British colonies might be overrun by other European 'races' (ibid.). As I put it, while the biological 'inferiority' of Britain's imperial subjects was perceived second-hand, the indigenous racism of the period was anti-Irish and anti-Semitic (e.g., Kirk, 1985; Miles, 1982). From the 1880s, there was a sizeable immigration of destitute Jewish people from the Russian pogroms, and this fuelled the preoccupation of politicians and commentators about the health of the nation, the fear of the degeneration of 'the race', and the subsequent threat to imperial and economic hegemony (Holmes, 1979; Thane, 1982). Intentional and overt institutional racism was rampant in all the major institutions of society: in the Government, the TUC and, of course, at the heart of capitalism itself (Cole, 2004a, p. 40). Moreover, racialization ensured that the institutionalized racism promulgated by the ruling class filtered down to the school and became part of popular culture. This was also most marked in the actual curriculum (for an analysis, see Cole and Blair, 2006, pp. 73–75; see also Cole, 1992b, pp. 71–80).

As I argued (Cole, 2004a, pp. 42–43), the Empire came home to roost after World War 2. The demands of an expanding post-war economy meant that Britain, like most other European countries, was faced with a major shortage of labor (Castles and Kosack, 1985). The overwhelming majority of migrants who came to Britain were from the Republic of Ireland, the Indian subcontinent and the Caribbean (Miles, 1989). Those industries where the demand for labor was greatest actively recruited Asian, black and other minority ethnic workers in their home countries (Fryer, 1984; Ramdin, 1987). Despite the heterogeneous class structure of the migrating populations (see

Heath and Ridge, 1983), migrant workers came to occupy, overwhelmingly, the semi-skilled and unskilled positions in the English labor market (Daniel, 1968: Smith, 1977). Furthermore, as I noted (Cole, 2004a, p. 42), they found themselves disproportionately concentrated in certain types of manual work characterized by a shortage of labor, shift working, unsocial hours, low pay and an unpleasant working environment (Smith, 1977). The consequences of this process of racialization were clear. According to Miles (1982, p. 165), these different racialized groups came to:

> occupy a structurally distinct position in the economic, political and ideological relations of British capitalism, but within the boundary of the working class. They therefore constitute a fraction of the working class, one that can be identified as a racialised fraction.

When the children of these migrant workers entered the education system, there were different kinds of representation for different minority ethnic group students (Cole, 2004a, pp. 43–47). While black children were seen as being disruptive and violent (corresponding to the previous racialization of their forebears in the African and Caribbean colonies), those of Asian background (namely Indian, Pakistani and Bangladeshi) were both seen as an academic (and social) threat to white children (see Hiro, 1971), or religious 'aliens' whose 'specific needs' posed a threat to the autonomy of schools (Blair, 1994)—a form of cultural racism. Asian students were also presented in seemingly benign terms as passive and studious and not presenting a disciplinary problem for teachers—a seemingly positive attribute (Cole, 2004a, p. 45). This notion of the 'passive Asian' student was juxtaposed against the 'aggressive' student of Caribbean origin and became, as Sally Tomlinson (1984) declared, 'a stick to beat the West Indian pupil with'. Racist inequalities in the *current* UK education system are discussed briefly in chapter 5 of this volume; see also Gillborn, 2008, Chapter 3 for a thorough discussion of these inequalities written from a CRT perspective.

The New Racial Domain in the United States

With respect to the United States, Manning Marable (2004) has used the concept of racialization to connect to modes of production there. He has described the current era in the United States as 'The New Racial Domain' (NRD). This New Racial Domain, he argues, is 'different from other earlier forms of racial domination, such as slavery, Jim Crow segregation, and ghettoization, or strict residential segregation, in several critical respects' (ibid.). These early forms of racialization, he goes on, were based primarily, if not exclusively, in the political economy of U.S. capitalism. 'Meaningful social reforms such as the Civil Rights Act of 1964 and the Voting Rights Act of 1965 were debated almost entirely within the context of America's expanding, domestic economy, and a background of Keynesian, welfare state public policies' (ibid.). The political economy of the 'New Racial Domain', on the other hand, is driven and largely determined by the forces of transnational capitalism, and the public policies of state neoliberalism, which rests on an unholy trinity, or deadly triad, of structural barriers to a decent life (ibid.). 'These oppressive structures', he argues, 'are mass unemployment, mass incarceration, and mass disfranchisement', with each factor directly feeding and accelerating the others, 'creating an ever-widening circle of social disadvantage, poverty, and civil death, touching the lives of tens of millions of U.S. people' (ibid.). For Marable, adopting a Marxist perspective, '[t]he process begins at the point of production. For decades, U.S. corporations have been outsourcing millions of better-paying jobs outside the country. The class warfare against unions has led to a steep decline in the percentage of U.S. workers' (ibid.). As Marable concludes:

> Within whole U.S. urban neighborhoods losing virtually their entire economic manufacturing and industrial employment, and with neoliberal social policies in place cutting job training programs, welfare, and public housing, millions of Americans now exist in conditions that exceed the devastation of the Great Depression of the 1930s. In 2004, in New York's Central Harlem community, 50 percent of all black male adults were currently unemployed. When one considers that this figure does not count those black males who are in the military, or inside prisons, its truly amazing and depressing (ibid.).Moreover, the new jobs being generated for the most part lack the health benefits, pensions, and wages that manufacturing and industrial employment once offered (ibid.).

Connecting to capitalist modes of production, for Marxists, is not as Mills (forthcoming, 2009) claims 'a manifestation of dogma', but a serious attempt to understand racism's interconnections with capitalism historically and contemporaneously. In making connections with modes of production, the Marxist concept of racialization, I must conclude, provides a more convincing account of racism in capitalist societies than do CRT emphases on 'white supremacy' on the one hand, and 'race' rather than class on the other.

Xeno-racialization

I have described the process by which refugees, asylum-seekers[23] and migrants to the United Kingdom from the newly joined countries of the European Union become falsely categorized as belonging to distinct 'races' as xeno-racialization (for an analysis, see Cole, 2004b, 2008c, 2008d, Chapter 9).[24] With respect to the EU's current enlargement, connections can be made between the respective roles of (ex-)imperial citizens in the immediate post World War II period, and migrant workers from Eastern Europe today (both sources of cheap labor). In addition, there are, as I have indicated, similarities in perceptions and treatment, something that is promoted by sections of the racist capitalist media.

The existence of xeno-racialization, although he did not use that term, along with other forms of racism, was recognized by the Chair of the

Commission for Equality and Human Rights in 2005, Trevor Phillips, when he noted:

> The nature of racism is changing subtly, but critically. We cannot respond by recycling the slogans of the '70s and '80s when race was regarded as a black and white affair. Today, we know that the reality of multi-ethnic, multi-faith Britain is more complex. Now, when we talk 'racial equality' and 'disadvantage', we are not necessarily referring to the needs of young black men. Rather we are speaking of the stigmatised [E]astern European asylum seeker; the Iraqi woman trapped in her own home by stone-throwing jobs; the Gypsies and Travellers who will live for 12 years less than the rest of us; and the Muslims unjustly victimised for atrocities committed by a tiny minority of followers of their faith…A recent…survey…shows that blatant discrimination or gross harassment is not found as frequently as in the past. But increasingly we are seeing the emergence of some other forms of racial bias which demand different tools (redhotcurry.com, 2005).

While CRT analysis serves as a constant reminder that racism is central in sustaining the current world order, the CRT concept of 'race-ing' (Crenshaw et al., 1995, p. xxvi), unlike the Marxist concepts of racialization and xeno-racialization, does not need to make the interconnections with modes of production since 'race' is itself material. In other words, oppression on grounds of 'race' can be explained merely as the modus operandi of 'white supremacy', a power structure in its own right.

To reiterate, I would argue that, in articulating with modes of production, these Marxist concepts of racialization and xeno-racialization have more purchase in explaining and understanding contemporary racism than 'white supremacy'. Indeed, I would maintain that if social class and capitalism are not central to the analysis, explanations are ambiguous and partial. Capitalism and social class are addressed in chapter 6 of this volume.

In this chapter I began by critiquing two of CRT's central tenets, the concepts of 'white supremacy' and the belief in 'race' as primary. I then outlined the definition of racism preferred by Marxist theorist, Robert Miles and his colleagues (a narrow one) before developing my own definition, which I argued, contra Miles, should be wide-ranging, finding this more useful than 'white supremacy' in understanding the multiple manifestations of racism in the contemporary neoliberal capitalist world. I then went on to make the case that the Marxist concepts of racialization and xeno-racialization have most purchase in explaining the processes by which certain groups become racialized at different phases in the capitalist mode of production. I will revisit the concept of racialization with respect to U.S. imperialism in chapter 6 of this volume. Having identified what I perceive to be CRT's two major weaknesses, in the next chapter I turn to what I perceive to be some of its strengths, strengths that nevertheless can be enhanced by Marxist analysis.

Chapter 3

The Strengths of CRT

I have argued that CRT is fundamentally flawed with respect to its advocacy of 'white supremacy', as a generally applicable descriptor of racism in certain modern societies, and in its prioritizing of 'race' over social class. In this chapter, I will assess what I perceive to be some of CRT's strengths. I will look specifically at *the use of the concept of property to explain historically segregation and white supremacy in the United States; the importance of voice; the concept of chronicle; the all-pervasive existence of racism in the world; interest convergence theory; contradiction-closing case transposition* and *CRT and the law in the United States.* I will indicate where I think these strengths can be enhanced by Marxist analysis.

It is a common misconception of many that Marxists argue that the economy determines everything; that nothing else needs to be taken into account; that Marxism draws on nothing other than the workings of the economy. Engels, in fact, challenged this economic determinist view of Marxism over one hundred years ago:

> According to the materialist conception of history, the *ultimately* determining element in history is the production and reproduction of real life. Marx and I have never said more than this.[1] Hence if somebody twists this into saying that the economic element is the *only* determining one, he transforms that proposition into a meaningless, abstract, senseless phrase. (Engels, 1895)

Indeed, prescient to CRT, Engels was well aware of the law and of ways that legal processes, including internal struggles therein, can influence the course of history. As he put it:

> political forms of the class struggle and its results… *juridical forms, and even the reflexes of all these actual struggles in the brains of the participants* [my emphasis], political, juristic, philosophical theories, religious views and their further development into systems of dogmas—also exercise their influence upon the course of the historical struggles and in many cases preponderate in determining their *form* (original emphasis). (Engels, 1890)

The use of the concept of property to explain historically segregation and white supremacy[2] in the United States.

Jessica DeCuir-Gunby (2006, pp. 101–105) gives the example of the case of Josephine DeCuir in the 1870s, a wealthy Creole[3] woman who challenged segregation, as practiced on the steamboat *Governor Allen*. DeCuir-Gunby (2006, p. 102) distinguishes four 'property rights of whiteness'; *rights of disposition (transferability); rights to use and enjoyment; reputation; and right to exclude.*

The *rights of disposition (transferability)* refer to the right to transfer 'whiteness'. Thus, although DuCuir was not black, neither was she completely white. Indeed 'her lack of completely white heritage made her Negro' (p. 102). This is because of the 'one drop rule' that declared any person with one drop of 'Negro' blood to be 'Negro' regardless of her or his physical appearance (p. 96). As DeCuir-Gunby (2006, p. 96) explains, '[a]ccording to this white supremacist belief system, the "one drop" of Negro blood made such a person contaminated and inferior, unworthy of being labelled white'. Creoles were thus 'negroes' who were given some rights afforded by whiteness, but not others, including the right to travel in the upper cabin of the *Governor Allen* (pp. 102–103).

The upper cabin of the ship represented the *rights to use and enjoyment* of a superior location. The upper cabin cost $2 more than 'the Negro section' (p. 103). As DeCuir-Gunby (ibid.) points out, this is significant, since charging white passengers more meant that they would be entitled to a better steamboat experience as well as better accommodation. Charging a higher price made it appear that white passengers were purchasing better services. However, given that 'negroes' were not permitted to pay this higher price for a better ticket, '[w]hite passengers were benefiting from their right to use and enjoyment based upon their whiteness' (ibid.). It meant in practice that a poor working class white person (provided s/he could get hold of the extra $2) had more right to utilize the upper cabin than a wealthy Creole person (p. 104).

With respect to the right of *reputation*, white passengers would go to great lengths to avoid 'racial' mixing and to ensure the reputation of whiteness. As John Benson, the boat's captain put it, '[i]f any boat was to attempt to mix the white and colored persons in the same cabin, I believe they would lose the white travel altogether' (*DeCuir v. Benson,* 1875, p. 73, cited in DeCuir-Gunby (p. 104)).

The final form of property right is the *right to exclude*. Benson provides the essence of this exclusion:

> The custom is...to give colored persons accommodation by themselves in a cabin appropriated to them exclusively and where the boats have no such accommodation it is customary to give them rooms by themselves or with the colored employees of the boat and giving them their meals after the white cabin passengers are through. (*DeCuir v. Benson,* 1875, p. 73, cited in DeCuir-Gunby (p. 105))

Thus the concept of property as used here is useful to explain historical events in pre–civil rights' United States.[4] Similar tales could be told of the United Kingdom before 'Race Relations' legislation, and, of course, of exceptional forms of the capitalist state (Poulantzas, 1978, p. 123) such as Nazi Germany and Apartheid South Africa.

The Importance of *Voice*

Delgado (1995) argues that the stories of people of color come from a different frame of reference, one underpinned by racism, and that this therefore gives them a voice that is different from the dominant culture and deserves to be heard. Arguing in a similar vein, Dixson and Rousseau (2006a, p. 35) define the concept of *voice* as 'the assertion and acknowledgement of the importance of the personal and community experiences of people of color as sources of knowledge'. Delgado and Stefancic (2001, p. 38) remind us that that Native Americans were great storytellers who used history and myth to preserve their culture and bind the groups together; that African Americans draw on a long history of storytelling that includes slave narratives, such as tales written to unmask the brutality behind the gentility promoted by the plantation society; and that in Latino society 'picaresque novelists made sly fun of social convention, puffed-up nobility, and illegitimate authority' (ibid.). Such 'stories' are important and need to be listened to in order to counter hegemonic discourse. Ladson-Billings (2006, p. xi) insists that CRT scholars are not 'making up stories', but constructing narratives 'out of the historical, socio-cultural and political realities of their lives and those of people of color' (ibid.). Moreover, as Ladson-Billings and Tate (1995, p. 58) argue, to make linkages between CRT and education, we need the voice of people of color to complete an analysis of the education system. Without the authentic voices of people of color (teachers, parents, administrators, students and community members, they conclude, 'it is doubtful that we can say or know anything useful about education in their communities' (ibid.).

While such insights are extremely illuminating, particularly for white people whose life experiences are restricted to monocultural settings, the crucial point for Marxists is that people of color need always to be listened to *because* they have been racialized in class societies. Racism can be best understood by both listening to and/or learning about the life histories and experiences of those at the receiving end of racism, and by objective Marxist analysis, which makes links with class and capitalism. There is thus considerable purchase in Zeus Leonardo's (2004) attempt to 'integrate Marxist objectivism and race theory's focus on subjectivity' (see also Maisuria, 2006). At the beginning of an article on racism in Britain, Alpesh Maisuria (2006, p. 1) explains his theoretical technique of linking state policy with his family's experiences of racism:

I will…[highlight] events and legislation that have shaped and defined macro policy, and also the micro experiences of the Maisuria family. It is of huge

important to establish a connection between macro politics and micro struggles in a liberal democracy to see how the state links with lived lives.[5]

As Steven Watts (1991, p. 652, cited in Darder and Torres, 2004,
p. 102) puts it, scholars who favor the use of narrative often 'fail to challenge the underlying socioeconomic, political and cultural structures that
have excluded these groups to begin with and have sustained the illusion of
choice'. Thus, as Darder and Torres, 2004, p. 102, point out, 'the narrative
and storytelling approach can render the scholarship antidialectical by creating a false dichotomy between objectivity and subjectivity', 'forgetting that
one is implied in the other, [while ignoring] a basic dialectical principle: that
men and women make history, but not in circumstances of their own choos-
ing' (Viotti da Costa, 2001, p. 20, cited in Darder and Torres, 2004, p. 102)
(the last phrase is, of course, adapted from Marx).

The Concept of Chronicle

Allied to *the importance of voice* is *the concept of chronicle*. Ladson-Billings
(2006, pp. viii–xi) gives an excellent example of a CRT 'chronicle'—a constructed narrative in which evidence and other forms of data are embedded.
Her chronicle concerns aggressive urban renewal and explains how city politicians and corporate leaders collude to exploit poor people of color in a U.S.
city, and how they connive to justify their actions. First the homeless are
forcefully removed; then tax breaks and incentives are introduced to bring
business back who will in turn hire people on low incomes and on limited
and part-time bases; then the older homes are destroyed, forcing their occu-
pants to move to the next poor community in order to make way for luxury
apartments. Finally, a tough law-and-order approach and a zero tolerance
policy on discipline and a rigid testing regime is applied to the schools so
that segregation appears in individual schools: for example, the privileged
white middle class children on the top floor of the school, and poor children
of color underneath. The chronicle ends with an excited interjection from
the construction firm owner:

> Yeah! This could work. And, when these kids can't pass tests and drop out of
> school we can hire them into those low-paying jobs we talked about. If they
> don't want those jobs and start doing anti-social things like drug dealing or
> stealing we can make a pitch for more and bigger prisons. I can build state-of-
> the-art super prisons in the suburbs. That will provide a steady stream of state
> employment for the White working class.

As I have argued the Marxist concept of racialization is important in establishing how groups of people become racialized at various historical con-
junctures. Thus Ladson-Billing's account above describes a typical example
of racialized capitalism in the United States in the twenty-first century.[6] In
this case, it involves people of color, but it could, at other stages in capitalist

development (in other countries) those on the receiving end could be Jewish people or Irish people, or Eastern European migrant workers. Chronicles are not the sole preserve of Critical Race Theorists, and in the Appendix to this chapter, I give an example of a Chronicle, grounded not in CRT, but subversive of one of its defining features, 'white supremacy'.

The All-Pervasive Existence of Racism in the World

It is tautologous to state that keeping racism at the forefront of analysis reminds us of the extent of racism in the world, and, as I have argued, to fully understand why this is the case, it is necessary to employ the Marxist concept of (xeno-) racialization, and to consider variations in (xeno-) racialized (and gendered) local, national and global capitalism and imperialism. Nevertheless, Critical Race Theorists highlight a number of pertinent points, which although needing to be analyzed using the above Marxist concepts, might, *without the intervention of Critical Race Theorists,* not be given the prominence they deserve. Thus it is important to point out that white-collar and corporate/industrial crime—perpetuated mostly by white people causes more personal injury, death and property loss than street crime, even on a per capita basis (Delgado and Stefancic, 2001, p. 43); to remind us about the number of black men in the United States incarcerated (more in prison than in college) (Delgado and Stefancic, 2001, p. 113) and having the death penalty disproportionately imposed; to highlight racial profiling both in the United States and in the United Kingdom; to report on how national assessment mechanisms in the United Kingdom actually *produce* inequality for black pupils (convincingly argued by Gillborn, 2006b; see also Gillborn, 2008, discussed in chapter 5 of this volume) even if such facts need relating to racialized capitalism. Critical Race Theorists are right that racism is not abnormal but inherent in the structures of (modern) societies. As Gillborn (2005, pp. 497–498) puts in a discussion about education, 'race inequity and racism are central features of the education system', and 'are not aberrant nor accidental phenomena that will be ironed out in time', but 'fundamental characteristics of the system'. In the same vein, Ladson-Billings (2006, p. xii) makes an important comment about Hurricane Katrina:

> Rather than fixate on the weather catastrophe and the breakdown of the social and political infrastructure, new CRT scholars in education look squarely at the way race was prefigured in the midst of the storm.

This, of course, is true. I would want to add though that, by and large, those who suffered were poor *working class* black people (Cole, 2006b). It needs to be pointed out as well, I would argue, that it was racialized capitalism that put them there in the first place. Scatamburlo-D'Annibale and McLaren (forthcoming, 2009) also point out how neoliberalism and the 'free market' lent itself to the utter destruction witnessed in New Orleans, in particular, lack of proper safeguards prior to the disaster.

Pronouncements by certain right-wing commentators served to mask this fact. As I have argued elsewhere (Cole, 2006b), the *Wall Street Journal* published a long editorial comment by Charles Murray, co-author of the aforementioned pseudo-scientific and racist tract, *The Bell Curve*. Entitled 'The Hallmark of the Underclass', conflating class and 'race', Murray declared that the hurricane merely demonstrated that 'the underclass has been growing during all the years that people were ignoring it' (cited in Van Auken, 2005). The images from New Orleans, he wrote, 'show us the face of the hard problem: those of the looters and thugs, and those of inert women doing nothing to help themselves or their children. They are the under-class' (cited in ibid.).[7] Murray also delved into a favorite topic of right-wing ideologues and pseudo-moralists like himself and Bennett—the 'illegitimacy rate' among blacks and 'low-income groups' generally (cited in ibid.). As Bill Van Auken concludes, underlining the extent of polarization in the United States:

> The reality is that Hurricane Katrina exposed the crisis and decay of an entire social system based on private profit and the accumulation of personal wealth at the expense of society as a whole. It likewise laid bare the immense social polarization between wealth and poverty in America—a chasm that has wid-ened over the course of decades. These grim social and class realities have inescapable revolutionary implications that have not been lost on America's ruling plutocracy. Its response will not be one of renewed social reformism or increased concern for a new generation of 'forgotten Americans'. On the contrary, it is turning even more sharply to the right, embracing the most nox-iously reactionary ideologies and relying ever more heavily on the police and military powers of the state. The resurgence of such fascistic conceptions as those of Bennett and Murray in the wake of Hurricane Katrina's devastation constitutes a grave warning to the American people. (ibid.)

The sickness of the U.S. capitalist order is perhaps exemplified by the fact that Bush rejected help with Katrina from both Venezuela and Cuba, the latter with 2,000 doctors ready with their equipment to go and save lives. (Chávez, 2008, cited in Campbell, 2008, pp. 59–60)

Thus CRT plays a useful role in keeping the all-pervasive existence of rac-ism in the world firmly at the forefront of their agenda. However, Marxist analysis is needed with this example of the resurgence of fascistic opinions and actions in the wake of Katrina, as with other examples, to analyze and to *understand* the material and ideological workings of racialized capitalism.

Interest Convergence Theory

Interest Convergence is a concept pioneered by Derrick Bell. It has two parts. First, 'the interest of blacks in achieving racial equality will be accommodated only when that interest converges with the interests of whites in policy-making decisions' (Bell, 2004, p. 69). Second, 'even when the interest con-vergence results in an effective racial remedy, that remedy will be abrogated at the point that policy makers fear the remedial policy is threatening the superior societal status of whites' (ibid.).

Following Bell (1980), Delgado and Stefancic (2001, p. 19) give the following example of the first form of Interest Convergence. *Brown v. Board of Education* in 1954 ruled that segregation based on color violated the Equal Protection Clause of the Fourteenth Amendment of the United States Constitution. Bell hypothesized that why this happened was that, given the fact that the Second World War had not long ended and the Korean War had ended in 1953, the possibility of mass domestic unrest loomed if African American service personnel, who had featured prominently in both wars, were subject to violent racism in the United States. Moreover, this period was the height of the Cold War, with much of the 'developing world' (much of it black, brown or Asian), uncommitted and up for grabs. As Delgado and Stefancic (ibid.) explain:

> It would ill serve the U.S. interest if the world press continued to carry stories of lynchings, racist sheriffs, or murders like that of Emmett Till [a fourteen year old African American boy who was shot, beaten and had his eye gouged out before being thrown into the river with a weight tied to his neck with barbed wire]. It was time for the United States to soften its stance toward domestic minorities. The interests of whites and blacks, for a brief moment, converged.

'U.S. interest' is, from a Marxist perspective, the interests of the U.S. racialized capitalist state. Moreover, I would argue that it is more accurate to describe 'whites in policy-making decisions' as 'the (white) racialized capitalist local or national state' and 'the superior societal status of whites' as 'the racialized status quo' in that society.[8]

Another example of the first form of Interest Convergence is given by Ladson-Billings (2005, p. 58), when in the State of Arizona, the governor argued that the state could not afford to observe the holiday for Martin Luther King, Jr. Day. However, after threatened boycotts from tourists, various African American civil rights groups, and the National Basketball Association, the state reversed its decision. As Ladson-Billings (ibid.) concludes, 'it did not have a change of heart about the significance of honoring Martin Luther King, Jr.; rather, the potential loss of revenue meant that the state had to have its interests converge with that of African Americans'. To this I would add that the decision reversal, of course, relates directly to the pursuit of surplus value by Arizonian capitalists at the time, supported by the racialized capitalist local state.

Following Morris (2001), Dixson and Rousseau (2006a, p. 44) give an instance which demonstrates both forms of Interest Convergence. Following *Brown v. Board of Education*, the St. Louis desegregation plan offered African American students the option of attending schools in the surrounding, predominantly white county districts. At the same time white students were enticed to return to city schools. Whereas many African American students took up the offer, far fewer white students did. The first form of Interest Convergence is demonstrated by the fact that the white county schools gained by having an overall increase in revenue; the second form of Interest Convergence by the failure to draw large numbers of whites back to the inner city, which would have entailed a threat to the social status of whites. In Morris's (2001, p. 592, words, cited in Dixson and Rousseau (2006a, p. 45), '[f]or [the white] parents, racial balance and equality are secondary to ensuring a quality education for their children'.

A Marxist perspective would be that here we have the largest metropolitan area in Missouri being forced to accommodate African American students and cashing in on the process, and the racist parents/carers of the white students, and no doubt the students themselves *because of their successful interpellation in a racist capitalist society* not wishing to mix with the racialized minority of African Americans, for fear of a drop in the standard of their education.[9]

Contradiction-Closing Cases

Gillborn (2008, p. 32) points out that there is a related but less well-known concept, also initially coined by Derrick Bell, 'the idea of the contradiction-closing case'. This is identified as 'those situations where an inequity becomes so visible and/or so large that the present situation threatens to become unsustainable' (ibid.). As Gillborn states, herein lies one of the dangers of interest convergence standing alone (p. 33). As he explains:

> While landmark cases may appear to advance the cause of justice, opponents re-double their efforts and overall little or nothing changes; except...that the landmark case becomes a rhetorical weapon to be used against further claims in the future. (ibid.)

Gillborn (ibid.) cites Delgado 1998, p. 445:

> Contradiction-closing cases...allow business as usual to go on even more smoothly than before, because now we can point to the exceptional case and say, 'See, our system is really fair and just. See what we just did for minorities or the poor'.[10]

Gillborn (2008, p. 33) concludes that *Brown v. Board of Education* provides a powerful example of contradiction-closing in that over fifty years on, even more African Americans attend schools that are de facto segregated. However, I would argue that this fact needs to be explained in the context of the continued and sustained racialization of U.S. capitalist society.

The Stephen Lawrence Case

The 'supreme example of a contradiction-closing case in the United Kingdom' is, for Gillborn (2008, p. 120) the Stephen Lawrence case. Gillborn

(2008, p. 132) is right to describe this case as 'one of the single most important episodes in the history of British race relations', so it is worthwhile
dwelling on this case for a while. The Stephen Lawrence Inquiry Report
(Macpherson, 1999) followed a lengthy public campaign initiated by the
parents of black teenager Stephen Lawrence, murdered by racist thugs in
1993. A bungled police investigation means that there have been no convictions. The Report looked at racism in the Metropolitan Police and other
British institutions, and acknowledged the existence of institutional racism
in the police, the education system, and other institutions in the society
(see chapter 5 of this volume for a discussion of institutional racism). While
there was a massive public outcry following the Report, with its findings
dominating the media, this was followed within weeks by skeptical comments from teachers union's leaders (Gillborn, 2008, p. 126). There were
also signs that the Education Department was not keen on pushing forward
the recommendations of the Inquiry. While it formally claimed to accept
the Report, it asserted that things were already in place to sort things out,
with the then Education Secretary, David Blunkett, claiming that the subject of Citizenship was already there to 'help children learn how to grow up
in a society that cares and to have real equality of opportunity for all' (cited
in Gillborn, 2008, p. 127).

Later, as Home Secretary, Blunkett was to suggest that 'institutional racism' was a slogan that let individual managers 'off the hook' in tackling
racism (Travis, 2003). He said that it was important that the government's
'diversity agenda' tackled the fight against prejudice but also took on the
long-standing need to change attitudes:

> That is why I was so worried about people talking about institutional rac-
> ism because it isn't institutions. It is patterns of work and processes that have
> grown up. It's people that make the difference. (ibid.)

Questioned about his comments afterwards, Blunkett added: 'I think the
slogan created a year or two ago about institutional racism missed the point.
It's not the structures created in the past but the processes to change structures in the future and it is individuals at all levels who do that' (ibid.).

As Gillborn (2008, pp. 130–131) explains, two significant events with
respect to the abandonment by the state of the concept of 'institutional racism' happened late in 2006, and early in 2007. First, in December, 2006, the
contents of an internal Education Department review were published in the
Independent on Sunday newspaper which specified 'institutional racism' as
the cause of black over-representation in exclusions from schools (Gillborn,
2008, p. 130). The review also warned that '[i]f we choose to use the term
"institutional racism", we need to be sensitive to the likely reception by
schools [but] if we choose not to use the term, we need to make sure that
the tone of our message remains sufficiently challenging' (cited in Gillborn,
2008, p. 130). The relevant minister, Lord Adonis, was quoted in the same
newspaper as arguing that 'since the report does not baldly conclude that

Britain's entire school system is "institutionally racist", the term—and the issue—could be quietly shelved' (cited in Gillborn, 2008, p. 144).

The second event happened early in 2007. The BBC reported that 'the Department of Health now regards the term [institutional racism] as "unhelpful" and believes that "the solutions lie in the hands of individuals not institutions" ' (cited in Gillborn, 2008, p. 131).

Gillborn (2008, p. 131) notes that 'by 2007 "institutional racism" had been erased from the policy lexicon'. A final blow to state endorsement of the concept came when the Commission on Equality and Human Rights (CEHR) (a body which oversees all equality issues) replaced the Commission for Racial Equality (CRE) and the principle of 'proportionality' was introduced. This allows public authorities not to take any action 'which might be disproportionate to the benefits the action would deliver' (cited in Gillborn, pp. 131–132). As Gillborn (2008, p. 132) points out, this means that schools can now decide to focus on other issues (e.g., underachieving boys). A school could acknowledge persistent 'race' inequalities, but decide that the effort required to reduce them was out of proportion to any benefits, or it could take no action on 'race' because it had too few minority ethnic students (ibid.). Gillborn (ibid.) concludes that the 'proposals signal a clear end to the period where equalities policy was drawn up with any meaningful reference to the Lawrence Inquiry'.

'The mere fact', Gillborn (2008, p. 133) concludes 'of the Stephen Lawrence and David Bennett Inquiries [the Bennett Inquiry on institutional racism in the National Health Service]—and the attendant press coverage—is assumed by some observers to denote change'. Herein lies the essence of a 'contradiction-closing case':

> This is exactly what Derrick Bell and Richard Delgado have warned about in relation to 'contradiction-closing cases'. The fact that institutional racism has been named explicitly as a factor in Britain's police, education and health services is not a solution, it is merely a diagnosis. (Gillborn, 2008, p. 133)

Gillborn (2008, p. 120) describes the Stephen Lawrence case as 'a case that, after years of the most painful campaigning and mistreatment, was supposed to have changed Britain for ever but...now seems to have left little imprint on the system in general, and education in particular'. However, he finishes on a positive note, arguing that the lesson from contradiction-closing cases 'is not that change is impossible but that change is always contested and every step forward be must be valued and protected. A victory won is not a victory secured' (p. 134). The whole case, he concludes, emphasizes 'the importance of constant vigilance to maintain and build upon each victory' (ibid.).[11]

No Marxist would disagree with this conclusion, nor with Gillborn when he insists that The Lawrence Inquiry was 'not granted by a benign state' that wanted to put right an injustice, but as a result of 'high profile protests and public demonstrations' (p. 135). As I have argued elsewhere

(Cole, 2008d, p. 75), unlike transmodernists for whom there is with oppressor a latent ethical potential (Dallmayr, 2004, p. 9), Marxists that all progressive gains for the working class have been gained by workers struggle. Marxists believe that, in essence, benefits, which accrue to workers in capitalist societies, have, in general, throughout history, been won by such struggle, rather than capitalist latent morality, or basic kindness. Indeed as Marx (1862, p. 1) put it, referring to the working class as the main source of extra-parliamentary pressure:

> No important innovation, no decisive measure has ever been carried through in this country without pressure from without, whether it was the opposition that required such pressure against the government or the government that required the pressure against the opposition. By pressure from without the Englishman understands great, extra-parliamentary popular demonstrations, which naturally cannot be staged without the lively participation of the working class.

The Case of Barack Obama

I would argue that mass popular resistance to George W. Bush as the leader of the U.S. Empire has created the possibility of a major instance of interest convergence and perhaps the beginning of a major contradiction-closing case in the United States. At the time of writing (Fall 2008) we have the distinct possibility of the United States' first black president. This can be used to 'prove' that the United States is no longer a racist society. As Angela Davis put it, though not specifying 'contradiction closure':

> [Obama] is being consumed as the embodiment of colour blindness. It's the notion that we have moved beyond racism by not taking race into account. That's what makes him conceivable as a presidential candidate. He's become the model of diversity in this period...a model of diversity as the difference that makes no difference. The change that brings no change. (cited in Young, 2008, p. 11)

Thus with its first black president, it might be more difficult to uphold charges of racism in U.S. society, imperial wars could continue and little else need to change, including the general thrust of Republican right-wing policies. Obama's policy on troop withdrawals from Iraq and on U.S. imperialism generally is ambiguous to say the least (Luce, 2008; Martin, 2008b; Van Auken, 2008c) As Bill Van Auken (2008a) explains, in speeches and press appearances, Obama continues to identify his campaign with support for American militarism, while at the same time backing away from his primary-campaign pledge to withdraw, on a definite timetable, U.S. combat forces from Iraq. As Obama put it in one speech: 'There is no challenge greater than the defense of our nation and our values' (ibid.). Obama went on to praise the actions of U.S. troops 'fighting a resurgent Taliban' and 'persevering in the deserts and cities of Iraq' (cited in ibid.). Backing away from his earlier pledge to carry out a 16-month withdrawal of combat troops from

Iraq, Obama stated, 'I have always said I would listen to the commanders on the ground. I have always said that the pace of withdrawal would be dictated by the safety and security of our troops and the need to maintain stability' (cited in ibid.). As the *Los Angeles Times* (September 11, 2008) put it:

> Beneath the harsh rhetoric, the two candidates—who meet today in New York City to commemorate the seventh anniversary of the Sept. 11 attacks—seem to be moving toward consensus on their broad-brush strategies. (cited in Van Auken, 2008c)

The newspaper went on to quote Brian Michael Jenkins, described as a 'leading authority on terrorism' at the Rand Corp. 'The process of political campaigning has exaggerated the differences of the two candidates on trivial issues', said Jenkins. 'But when it comes to where the campaigns have outlined their platforms on Iraq, Afghanistan and national security, there isn't a great deal of difference.' In other words, as Van Auken, 2008c concludes, 'both candidates support policies that translate into the protracted occupation of Iraq—albeit, if possible, with fewer ground troops—an escalation of the war in Afghanistan and its extension across the border into Pakistan' (see also Jones, 2008).

With respect to the domestic agenda, Gary Younge, (2008, p. 12) points out that Obama's policies on healthcare and the mortgage crisis were the least comprehensive of the main primary candidates, and his advisory team is full of neoliberals (see also Goldenberg, 2008). Moreover, since he became the Democratic Presidential Candidate, he has moved further to the right. This, however, needs to be understand in the context of U.S. capitalism, and, once again can be supplemented by Marxist analysis. As Van Auken puts it (2008c), '[t]he lurch to the right by the Obama campaign is so blatant that it has aroused substantial commentary in the bourgeois press'. Some of it, he goes on, is gloating and some of it reflects concerns that this maneuver is so blatant that 'it may alienate substantial layers of the population from the electoral process and expose the fraud of the entire two-party system'. Thus the *Wall Street Journal* in an article entitled 'Bush's Third Term' pointed to Obama's continuous warnings against McCain's victory resulting in 'George Bush's third term': 'Maybe he's worried that someone will notice that he's the candidate running for it'. Van Auken (2008c) concludes, underlining the fundamental need for this prime example of interest convergence and supreme indication of a contradiction-closing case to be related to the dynamics of U.S. capitalism:

> One thing is certain. The policies of an incoming Obama administration will not be determined by the erstwhile populist posturing of the candidate or by the pressure exerted by the left liberals. Rather, they will be dictated by the enormity of the economic and political crisis confronting American capitalism and what is required under these conditions to defend the class interests of the ruling elite. The turn to the right on the campaign trail is preparation for this essential task.

In summary then, the CRT concepts of *Interest Convergence* and *Contradiction-closing cases* are helpful in understanding economic and political processes, and are readily amenable to considerable enhancement by Marxist theory.

Transposition

'Transposition' is another useful CRT developed by Gillborn (2008, p. 82) and adapted from Gregg Beratan (forthcoming, 2008) (it was originally applied to disability issues). Transposition describes 'situations where one form of injustice is legitimized by reference to a different, more readily acceptable form of argument' (Gillborn, 2008, p. 82). As Beratan (forthcoming, 2008, cited in Gillborn, 2008, p. 82) notes, when a musical piece is transposed into a different key, the sound changes, but the song remains fundamentally the same. Gillborn (ibid.) gives the example of racist transposition in 'the strategic deployment of gender equity issues as an acceptable trope for otherwise aggressively racist attacks on Muslim communities'. Thus we have, in the aftermath of 7/7, Tony Blair (then Prime Minister) arguing against 'forced marriage', 'marriage at an early age' and women being debarred from certain mosques. While there is no doubt about the justice of these statements *with respect to women's rights*, the *intention* in making them is the issue here. The reason for Blair's intervention on behalf of Muslim women *at this time* needs to be interrogated. For Gillborn (2008, p. 84) 'Blair was keen to emphasize his (rhetorical) commitment to equity and... the act of transposition was in evidence as the attacks on Islam were presented as a concern for women's rights'. As Steve, a character in one of Gillborn's chronicles (see the section of this chapter on 'Chronicle'; see also the Appendix to this chapter) puts it, we have

> Tony Blair pouring out his heart about the importance of fighting sex discrimination while he supports new immigration rules that will make it harder for women to enter the country to be with their husbands.[12] And if they do ever get in, Gordon Brown wants them to do community service! (Gillborn, 2008, p. 15)[13]

The Professor, the other character in Gillborn's chronicle, responds that it is a very effective technique in that it splits the opposition, and allows policymakers to appear liberal and concerned when they are actually proposing measures that are more and more regressive (Gillborn, 2008, p. 15).

Like other CRT concepts discussed in this chapter, *interest convergence*, *contradiction-closing cases* and *transposition* do make a contribution to our understanding of racism and racist processes, but like the other concepts, they can only stand alone if one accepts the tenet of 'white supremacy'. From a Marxist standpoint, there is a need to situate *interest convergence*, *contradiction-closing cases* and *transposition* firmly in the context of the machinations of capitalism.

To take the Stephen Lawrence case, it is hardly likely that the racialized capitalist state is going to tolerate for long the notion that its major domestic Repressive State Apparatuses (the police) *and* its main Ideological State Apparatus (the education system) are institutionally racist.[14]

With respect to Obama, given the unpopularity of Bush, it converges with the interest of U.S. capitalism to have the possibility of a black president who will carry on business more or less as usual. However, Obama *is* an improvement on Bush and we should bear in mind Gillborn's point that the lesson from contradiction-closing cases is that change is always contested and every step forward be must be valued and protected, and that victories can be built upon.

As far as Blair's and Brown's rhetoric are concerned, this needs to be seen in the context of the legacy of old UK imperialism and UK-supported new U.S. imperialism (see chapter 6 of this volume; see also Cole, 2008d, pp. 98–111).

CRT and the Law in the United States

CRT has performed and continues to perform a useful reformist intervention in the U.S. legal system. As Crenshaw et al., 1995b, p. xxviii, explain, CRT tells us that the courts' claims to 'color-blindness' is in reality 'the product of a deeply politicized choice' rather than 'an ineluctable legal logic': '[t]he appeal to color-blindness can...be said to serve as part of an ideological strategy by which the current Court obscures its active role in sustaining hierarchies of racial power' (ibid.). Crenshaw et al. (ibid.) conclude that CRT 'offers a valuable conceptual compass for mapping the doctrinal mystifications which the current Court has developed to camouflage its conservative agenda'. As an example, Delgado and Stefancic (2001, p. 114–115) recount how one federal judge, versed in Critical Race Theory, and required under a 'three strikes and you're out' type law to give out a longer sentence to a black motorist, declined to do so on the grounds that black motorists tend to be pulled over more frequently than whites as a result of racial profiling, and that the defendant's two prior convictions had likely been tainted by racism.

Marxists would be totally supportive here. Indeed, Marxists support all reforms and ameliorations of injustice within capitalism, but with a longer term commitment to its overthrow and replacement with democratic socialism (see chapter 7 of this volume for a discussion).

In this chapter, I have discussed some of the strengths of CRT. However, I have also argued that these strengths need supplementing with Marxist analysis, in order for them to be contextualized. In the next chapter, I will discuss multicultural education and the antiracist response (based on Marxism) first in the United States, then in the United Kingdom.

Appendix

David Gillborn (2008, pp. 5–19; 184–196) gives examples of two Chronicles, which trace conversations between an older white professor and a young black male part-time law student. As Gillborn (2008, p. 4) explains Critical Race Theorists (e.g., Derrick Bell, Richard Delgado and Patricia Williams) tend to use as characters, usually a law professor of color and a younger alter-ego. In the Chronicle that follows I will use Gillborn's persona of the Professor. But instead of a young black male student, my other character is a young white female health care assistant, Jo. As in Gillborn's Chronicles, the Professor is not Gillborn, but shares 'some of [his] thoughts, fears and experiences but...[is] also free to say and think things that [Gillborn] would not necessarily support' (Gillborn, 2008, p. 5). My Chronicle attempts to subvert and question the validity of the CRT concept of 'white supremacy'.

Rather than the Harvard style of referencing used in the rest of this volume, in order that the conversation flows more naturally, in this appendix I have used endnotes only.

Chronicle: CRT, White Supremacy, and Racism

The Professor had just given the first of a series of public lectures on CRT, organized by the new Centre for Education and Social Justice at the University of the West Midlands.

'I enjoyed your lecture Professor Glindvale. Can I just ask you one question?'

The Professor smiled. 'Good. Yes, of course. And please call me "Rob"'.

'Ok, *Rob*. And I'm Jo'. The professor held out his hand. 'So what's the question?'

Jo glanced at the whiteboard which still had the professor's concluding slide, containing the words, 'The Perspectives of White people are constantly enforced over those of minoritized groups'.

Jo looked the Professor squarely in the eye. 'I find the CRT concept of "white supremacy" a problem. I don't think it applies to *all* white people'

The Professor frowned. He had been asked the same question many times before. 'Look, this doesn't mean that all white-identified people draw similar benefits but white people are all implicated in white supremacy, and they do all benefit, whether they like it or not...'

Jo interjected. 'The people on the Rimene estate here, the white working class estate, how does "white supremacy" relate to them? And what about the migrant workers from Eastern Europe? There are lots in the Midlands now'

The Professor looked at Jo in earnest. 'I recognize that some white people are subject to racism—Sivanandan has used the term "xeno-racism" to describe racism meted out to Eastern Europeans. However, to reiterate, while white-identified people are not all active in identical ways, and do not all draw similar advantages, we are all implicated in relations of shared power and dominance. We all benefit, whether we like it or not. In general we don't get followed by security people in supermarkets, we don't get hassled on racial grounds, we *can* act as we want without being seen as representing an entire racial group'[1]

At that point, the Head of the Centre for Education and Social Justice announced that everyone was welcome to carry on the discussion in a nearby pub, *The Ivy League Inn.*

The Professor picked up his large old leather bag that was his constant companion. 'I've got to sort out a few things. Are you going to the pub? Cos if you are, we can continue the conversation there?'

Jo glanced again at the slide on the whiteboard. 'Yes, that would be great'.

Just after half an hour later, Jo arrived in *The Ivy League Inn* to find the Professor and the Head of the Centre for Education and Social Justice in heated debate. The Professor caught her eye, and beckoned her over. 'What can I get you?'

'A pint of Bergermeister would be nice'. The Professor moved to the bar to buy the drink, and Jo looked at the Head of the Centre. 'I find your Marxist analyses of racism very interesting, particularly your work on xeno-racism and xeno-racialization. Do you know of any work on racism directed against other white people?'

The Head of the Centre looked pleased. 'CRT has its good points, but one of its many shortcomings is its inability to understand and to analyse non-color-coded racism. I'll email you some sources on this'.

At that point, the Professor returned with the beer.

The Head of the Centre looked at his watch. 'Rob, the discussion's been great as *always,* but I've got to get this train'.

The Head of the Centre shook hands with the Professor. 'Great to see you. You too, Jo. I'll be in touch'.

The Head of the Centre left the pub, and the Professor and Jo moved to a large table, where others who had attended his lecture were seated.

After a while people began to drift off. Jo turned to the professor. 'I've had a lot of thoughts about white privilege and power and dominance

And I think there's a lot I could say'.

The Professor gave Jo his card. 'Ok, you can reach me here'.

Jo nodded in agreement. 'Now before you go, I've got something for *you* to read. It's stuff my partner jotted down in *The St. George Tavern,* where she works. You see *The Ivy League* isn't at all typical of pubs in this town'.

The Professor put the A4 sheet of paper in his bag. 'Well, you'll email me then? Good to meet you'.

Jo nodded. 'Good to meet you too'.

Twenty minutes later, the Professor was on the slow train to London. He opened his brief case and took out the crumpled sheet of A4, headed *St. George Tavern Tuesday,* and read the following:

'I had gallstones, but they got rid of the c***s. Have to go back. Problem is the NHS is full of f***ing ni**ers these days'[2]

Taxi driver enters the bar. As he picks up customer's bag, customer makes a 'joke' about having a bomb in it. Taxi driver responds, 'I'll drop the bomb in the nearest mosque. Anyone object to that? C'mon anyone object to that? C'mon let's hear it'. [general sounds of approval]. 'Nobody. Right?'[3]

The Professor looked alarmed, nodded and thought to himself 'white supremacy in action. It'll be interesting to see what Jo has to say about white privilege, power and dominance'.

Back in London, the following day, the Professor opened the door to his office. He threw his bag across the room, where it landed perfectly against his desk in the usual place. He turned on this computer. There was an email from Jo:

> I stayed up all night reading and thinking about our conversation, and I came to the conclusion that white privilege doesn't apply to me in any way at all. Nor does power and dominance.

Astounded, the Professor turned his kettle on, and continued to read Jo's e-mail:

> I do get followed by Security when I'm in the shops.

The Professor was growing extremely curious. He glanced at his watch, and realized that he had a departmental meeting in five minutes. He added a spoonful of instant coffee to his old mug, filled it with water, and headed, cup in hand, off to the meeting.

An hour later, he rushed to his room, eager to read the rest of Jo's email. He noticed straight away, however, that there was a second email from Jo. He opened it, noting that it was sent to the Head of the Centre for Education and Social Justice, and copied to him. The Professor read the first few lines:

> Thanks for the stuff you emailed me re racism directed at white people. I would like to have a look at Mac an Ghaill's book[4] on anti-Irish racism...

The Professor closed this email, and re-opened the first one from Jo.

> I stayed up all night reading and thinking about our conversation, and I came to the conclusion that white privilege doesn't apply to me in any way at all. Nor does power and dominance. I do get followed by Security when I'm in the shops. The old stereotypes apply. You see I got a lot of publicity when I stood for the Council a few months back, including photos in the press, and even an appearance on local TV. I do get harassed a lot of the time, and am seen as part of a particular group. And that's a lot to do with why I didn't get elected.

The Professor was getting more and more curious. He decided to have a look at the other email from Jo:

> Thanks for the stuff you emailed me re racism directed at white people. I would like to have a look at Mac an Ghaill's book on anti-Irish racism, and thanks for offering to lend me a copy. Can you bring it to the next *Respect* meeting, if you're coming?[5] More racist crap in the *St. George* for your new book. This guy was talking about that accident the other day and spouted

out: 'They're all fucking asylum-seekers. They don't know our traffic rules'.[6] I totally agree with Louis Proyect[7] that 'the culture of the Roma people is about as at odds with the profit-driven world of real estate and banking as can be imagined'. He also makes interesting comments about Roma people being enslaved in Romania and being murdered by the Nazis. He's also right about not getting the moral and financial compensation that Jewish people have quite rightly got. He's also right about some of the world's greatest musicians. He's right that we are still despised and persecuted by many as you point out in your book, when you talk about *The Sun*.[8] Finally, he is right that we have never made war or invaded another country. The fact that I have white skin is neither here nor there. To the racists, whether we identify as English Romany or Irish Romany, or gypsy or traveller, or gypsy/traveller, or whatever, we in the GRT communities are all pikies, and we get all the usual stereotypes. Great that that awful word is now officially a 'race' hate word[9] . . .

The Professor browed his forehead. He turned on the kettle, and picked up a copy of his latest book from the small coffee table, which as always was covered in papers and computer disks. He flicked through the index of the book, looking for the sections on 'whiteness' 'white privilege' and 'white supremacy'. He emptied the last granules of coffee from the jar into his cup, and added the boiling water. He was going to have to think very seriously about how he was going to respond to Jo.[10]

Notes

1. Gillborn, 2008, pp. 34–35.
2. This is the transcript of part of a real conversation; see Cole, 2008c, p. 13.
3. This was also really part of a real conversation; see Cole, 2008c, p. 13.
4. Mac an Ghaill (2000).
5. The platform of the UK socialist organization, *Respect* is referred to in chapter 7 of this volume. Unfortunately, it has recently split, and its future is uncertain.
6. This was also really said; see Cole, 2008c, p. 13.
7. (Proyect, 2007).
8. See p. 30 of this volume.
9. Following the conviction of Lee Coleman who repeatedly used the racist term, 'pikey' in a drunken outburst over a nightclub entry fee, 'pikey' has become a 'race' hate word (*Metro* December 13, 2007). During his outburst Coleman made stereotypical racist accusations of theft.
10. The fact that it does not occur to the Professor (convinced of the efficacy of CRT notions of 'white supremacy') that Jo might be a member of a *white* oppressed group, in this case a member of one of the GRT community groups, underlines the third of four problems with the concept of 'white supremacy' that I identified in chapter 2 of this volume, namely that 'white supremacy' inadequately explains non-color-coded racism.

Chapter 4

Multicultural Education and Antiracist Education in the United States and the United Kingdom

In the first part of this chapter, I begin by discussing three forms of reactionary multicultural education in the United States identified by Peter McLaren. I go on to analyze McLaren's advocacy, *in his postmodern phase,* of 'critical resistant multiculturalism', a form of multiculturalism favored by Critical Race Theorist Gloria Ladson-Billings. I conclude the section of the chapter on the United States by appraising McLaren's promotion, *since he returned to the Marxist problematic,* of 'revolutionary multiculturalism'.

In the second part of the chapter, after arguing that Britain is and always has been a multicultural society, I go on to discuss the differing ideological perspectives of multicultural and antiracist education respectively. In chapter 3, I discussed how antiracism, in particular the *temporary* state endorsement in the light of the Stephen Lawrence Inquiry Report (Macpherson, 1999) of 'institutional racism', has been undermined by subsequent 'official' denials. In this chapter, I argue that the gains accrued to the antiracist standpoint by the Inquiry are further under threat from New Labour's promotion of a hard version of 'community cohesion'.

Forms of Multiculturalism in the United States

Prior to the arrival of CRT, educational theory in the United States, when dealing with 'race', tended to center on the concept of 'multicultural education'. In a critique of certain reactionary forms of multicultural education in the United States, Peter McLaren (1995, p. 119) has identified three: conservative (or corporate) multiculturalism; liberal multiculturalism; and left-liberal multiculturalism. These, according to McLaren (ibid.) are 'ideal-typical labels meant to serve only as a "heuristic" device'. As such they are useful in conceptual analysis.

Conservative Multiculturalism

Conservative (or corporate) multiculturalism is about disavowing racism while upholding corporate power (Ladson-Billings, 2005, p. 53). As Ladson-Billings puts it, '[c]orporate or conservative multiculturalism has a veneer of diversity without any commitment to social justice or structural change' (p. 53). She gives the example of major U.S. retailer, Sears, who has targeted black and Latino/a consumers (2005, p. 50). Recalling how the *Chicago Tribune* headlined, 'A Multicultural State for Sears', she notes that the 'ease with which a major newspaper used the term *multicultural* tells us something about how power and domination appropriate even the most marginal voices' (p. 50), in order 'to promote consumption (and perhaps exploitation of workers)' (p. 53).

In terms of the school curriculum, Ladson-Billings (ibid.) likens conservative (or corporate) multiculturalism to J. E. King's (2001, p. 274) description of 'marginalizing knowledge'—a 'form of curriculum transformation that can include selected "multicultural" curriculum content [but] that simultaneously distorts both the historical and social reality that people actually experienced'. So even though there may be representations of minority groups in texts and school curriculum, they may well be marginalized (p. 53). Ladson-Billings (ibid.) cites a typical textbook strategy of placing information about subordinated groups in a 'special features' section while the main text, carrying the dominant discourse, remains 'uninterrupted and undisturbed by "multicultural information"'.

Liberal Multiculturalism

Liberal multiculturalism, in McLaren's (1995, p. 124) words, is based on the premise of 'intellectual "sameness" among the races, on their cognitive equivalence or the rationality imminent in all races that permits them to compete equally in a capitalist society'. In other words, it is predicated on meritocracy and equal opportunities within capitalism (Ladson-Billings, 2005, p. 53). Ladson-Billings (ibid.) once again utilizes a J. E. King concept, namely 'expanding knowledge' to locate this variety of multiculturalism. 'Expanding knowledge' refers to the process of multiculturalizing knowledge 'without changing fundamentally the norm of middle classness in the social framework's cultural model of being' (King, 2001, p. 275, cited in Ladson-Billings, 2005, p. 53). It tries to address the concerns of all groups equally without disturbing the power structure (Ladson-Billings, 2005, p. 53). Thus, as Ladson-Billings explains, most campuses offer program for all groups, but in isolation, and white middle class norms prevail (ibid.). Moreover, as she concludes, '[b]y acknowledging the existence of various groups while simultaneously ignoring the issues of power and structural inequity, liberal multiculturalism functions as a form of appeasement' (ibid.).

Left-Liberal Multiculturalism

McLaren's (1995, pp. 124–125) third form of multiculturalism, the 'left-liberal' variety, in Ladson-Billings' (2005, p. 54) words, exoticizes cultural differences. As McLaren (1994, p. 51, cited in Ladson-Billings, 2005, p. 54) puts it, this approach 'locates difference in a primeval past of cultural authenticity'. In an earlier paper with William Tate (Ladson-Billings and Tate, 1995, p. 61) this form of multicultural education is described as trivia: 'artefacts of cultures such as eating ethnic or cultural foods, singing songs or dancing, reading folktales' (as we shall see later in this chapter, the dominant form of multicultural education in the United Kingdom is very similar to this approach).

Critical and Resistant Multiculturalism

McLaren (1995, pp. 126–144) offers, in place of these three forms of multiculturalism, a notion of 'critical and resistant multiculturalism'. He describes this as 'a resistance post-structuralist approach to meaning...located within the larger context of postmodern theory' (p. 126). Critical and Resistance Multiculturalism 'stresses the central task of transforming the social, cultural, and institutional relations in which meanings are generated' (ibid.). Resistance multiculturalism 'doesn't see diversity itself as a goal but rather argues that diversity must be affirmed within a politics of cultural criticism and a commitment to social justice' (ibid.). The 'central theoretical position of critical multiculturalism' is 'that differences are produced according to the ideological production and reception of cultural signs' (p. 130). The politics of signification, McLaren goes on, are at work in the way minority students, unlike white middle class students, are considered for 'behavioral' placements (ibid.). A critically multiculturalist curriculum, McLaren (ibid.) suggests 'can help teachers explore the ways in which students are differentially subjected to ideological inscriptions and multiply-organized discourses of desire through a politics of signification'.

Like post-structural and postmodern analyses in educational theory in general, there is much talk of social change and social justice in McLaren's *mid 1990s* analysis, but no concrete suggestions for societal change. Elsewhere (Cole, 2008d, Chapter 5) I have examined the work of leading post-structuralists/postmodernists, Patti Lather, Elizabeth Atkinson and Judith Baxter. I argue that, while many questions are asked, and there are many claims for moving towards social change and social justice, no specific indications are given except at the local level.[1] One of the great strengths of Marxism is that it allows us to move beyond appearances and to look beneath the surface *and* to move forward collectively at local, nationally and internationally. It allows us to transgress Derrida's (1990, p. 963) 'ordeal of the undecidable', Lather's (2001, p. 184) 'praxis of not being so sure', which is itself derived from the 'ordeal of the undecideable', and Baxter's (2002) 'paralysis of practice' (see Cole, 2008d, Chapter 5).[2]

CRT and a Rights-Based Discourse

While favoring McLaren's concept of 'critical and resistant multiculturalism', Ladson-Billings (2005, pp. 54–57) nevertheless identifies some tensions within the field of multiculturalism *per se* and notes not only issues of feminism, sexuality and social class, but also religious dimensions. As she explains, (p. 57) '[w]hat one group perceives as the multicultural agenda is something else for another'. She concludes that the 'complexity of identities that individuals experience makes it difficult to craft a multicultural mission that speaks to the specificity of identity' (ibid.) and that 'attempts to be all things to all people seem to minimize the effective impact of multicultural education as a vehicle for school and social change' (ibid.).

She thus (ibid.) turns to CRT as a way forward, specifically its capacity to formulate 'a rights-based discourse'[3] CRT was deeply entrenched in legal scholarship until Derrick Bell (1992a) published *And We Are Not Saved.* As Ladson-Billings (2006, p. viii) notes, in Bell's book 'we see examples of disciplinary mergers that pull on the traditions of education, sociology, anthropology, music and art scholarship'. CRT was, in fact, introduced into education research by Ladson-Billings and William Tate at the 1994 American Education Research Association Annual Conference in New Orleans, and subsequently published as Ladson-Billings and Tate, 1995.

Ladson-Billings and Tate (1995, p. 59) apply their discussion of property rights—*rights of disposition, rights to use and enjoyment, reputation and status property the absolute right to exclude* to education. With respect to *rights of disposition*, Ladson-Billings and Tate (ibid.) argue that students are rewarded only for conformity to perceived 'white norms' or sanctioned for certain cultural practices, such as dress, speech patterns, unauthorized conceptions of knowledge.

Second, *rights to use and enjoyment* are differential: this can be material differences in the school, and unequal space, with white schools faring much better. In addition, the curriculum can be more conducive to deeper thinking skills in 'white schools' (ibid.).

As far as *reputation and status property* is concerned, to designate a school or program as 'nonwhite' is to diminish its reputation or status. Moreover, despite the prestige of foreign language learning in the United States, bilingual education as a nonwhite form of language learning has lower status (p. 60). The same is true of the differential status of suburban (white) and urban (nonwhite) schools.

Finally, in schooling, *the absolute right to exclude* was demonstrated initially by denying black children the right to schooling altogether (ibid.). This was modified later by segregated schooling. Ladson-Billings and Tate argue, it has also been demonstrated by 'white flight' (ibid.). Within schools *the absolute right to exclude* includes tracking (setting in UK terminology), and 'gifted' programs (ibid.).[4] 'So complete is this exclusion' they conclude, 'that black students often come to the university in the role of intruders—who have been granted special permission to be there' (ibid.).

Congruent with the CRT view that civil rights law is regularly subverted to benefit whites, they argue that multicultural reforms are routinely 'sucked back into the system' (p. 62.). Being 'mired in liberal ideology', they go on, 'the current multicultural paradigm...offers no radical change in the current order' (ibid.). Making it clear that they are not attacking the proponents of multicultural education, they note the near nigh impossibility of 'maintaining the spirit and intent of justice for the oppressed while simultaneously permitting the hegemonic rule of the oppressor' (ibid.).

Ladson-Billings (2005, p. 59), in discussing the curriculum, notes that 'CRT sees the official knowledge…of the school curriculum as a culturally specific artefact designed to maintain the current social order' (ibid.) which erases or 'sanitizes' people of color, women and anyone else who challenges the 'master script' (ibid.). The curriculum is also differentiated according to 'race' and class, with children of the dominant group having an 'enriched' and 'rigorous' curriculum (ibid.), and the subordinated group spending 'most of the day with no curriculum outside of test preparation' (ibid.). This restricted curriculum, Ladson-Billings (ibid.) following Cheryl Harris (1993) argues is a good example of use and enjoyment of property, with 'Whites' having 'rights of disposition, use and enjoyment, reputation and status—and the absolute right to exclude' (ibid.). This doesn't occur by chance. 'The infrastructure and networks of Whiteness provide differential access to the school curriculum' (ibid.).

Ladson-Billings (pp. 59–60) also addresses pedagogy, arguing, following Haberman, 1991, that there are two types of pedagogy, a 'pedagogy of poverty' serving poor urban students, and focusing on the basics, and 'good' teaching involving a much more in-depth engagement with important issues. Finally, she looks at assessment which legitimizes the 'deficiencies' of children of color, poor children, immigrants and limited English-speaking children (p. 60) and substantiate inequity and validate the privilege of those with cultural capital (ibid.).

While Marxists would challenge some of the terminology here—the privileging of 'whites' per se (see chapter 2 of this volume); 'networks of Whiteness' (without connections to social class), they would have little basic disagreement with Ladson-Billings' analysis of schooling in the United States. Indeed, Marxists have been arguing on similar lines since the publication of Bowles and Gintis's (1976) *Schooling in Capitalist America* (which of course related these issues firmly at the root of the problem: the capitalist economy), without the need to invoke Critical Race Theory.[5]

Despite her skepticism about multiculturalism, Ladson-Billings (2005, p. 62) nevertheless endorses McLaren's aforementioned post-structuralist concept of 'critical and resistant multiculturalism'.[6]

Revolutionary Multiculturalism

Having advocated 'critical and resistant multiculturalism' in the 1990s, McLaren has subsequently abandoned it in favor of revolutionary multiculturalism. This parallels his move from postmodernism to Marxism.[7] McLaren and Ramin Farahmandpur (McLaren and Farahmandur, 2005, p. 147) advocate revolutionary multiculturalism, as opposed to 'critical multicultural education', as a framework 'for developing a pedagogical praxis…[which] opens up social and political spaces for the oppressed to challenge on their own terms and in their own ways the various forms of class, race, and gender oppression that are reproduced by dominant social relations'. Scatamburlo-D'Annibale and McLaren (forthcoming, 2009) explain the move. They argue that:

> while racism often does take on a life of its own, its material basis can be traced to the means and relations of production within capitalist society—to the social division of labor that occurs when workers sell their labor-power for a wage to the capitalist (i.e., to the ownership of the means of production). To ignore class exploitation when you are talking about racism is a serious mistake.

Addressing the 'multiculturalist problematic' and following E. San Juan (2004), Scatamburlo-D'Annibale and McLaren (forthcoming, 2009) argue that the separation of 'race' and racism from the social relations of production effectively treats them mainly as issues of ethnicity and the politics of 'difference'. In doing so, they go on, this 'effectively neutralizes the perennial conflicts in the system by containing diversity in a common grid and selling diversity in order to preserve the ethnocentric paradigm of commodity relations that structure the experience of life-worlds within globalizing capitalism'.

The educational implications are that it is important to bring education into conversation with movements that speak to the larger totality of capitalist social relations and which challenge capital's social universe. The strategic focus for Marxist educators, if we are to have effective antiracist, anti-sexist, anti-homophobic struggles, should be, they argue, on capitalist exploitation (ibid.). Critical Race Theory and nonrevolutionary multiculturalism, they continue, fail to foreground the fundamental importance of the social division of labor in the capitalist production process as a key factor in understanding racism. 'The reason we need to focus on a critique of political economy in our anti-racist efforts', they argue, 'is that racism in capitalist society results from the racialization of the social relations of capitalist exploitation' (ibid.). The cite J. B. Foster, 2005:

> The various forms of non-class domination are so endemic to capitalist society, so much a part of its strategy of divide and conquer, that no progress can be made in overcoming class oppression without also fighting...these other social divisions

Ultimately, Scatamburlo-D'Annibale and McLaren (forthcoming, 2009) conclude that 'race' alone is too blunt an analytical tool to effectively elucidate complex social phenomena, even when, as with Katrina, social inequalities are expressed in such blatantly racial disparities. McLaren is unequivocal and his commitment to a class analysis, and to the struggle for socialism. As Suoranta McLaren and Jaramillo (forthcoming, 2009) put it:

> Our own attempts to develop a radical humanistic socialism—in part by de-writing socialism as a thing of the past—assumes the position that socialism and pedagogical socialist principles are not dead letters, but open pages in the

book of social and economic justice yet to be written or rewritten by people struggling to build a truly egalitarian social order.

The role of revolutionary educators is to:

> invite students to recollect the past, to situate the present socially, politically and economically…In this way, critical scholars can help students to challenge the particularities of their subjective existence in relation to the larger socio-cultural and economic frameworks that give them meaning, thereby contesting the erasure of their cultural and subjective formations while at the same time dialectically refashioning their self and social formations in their struggle to become the subject rather than the object of history. And this means remembering that the pedagogical is the political. And creating pedagogical spaces for self and social transformation, and for coming to understand that both are co-constitutive of building socialism for the twenty-first century—a revolutionary praxis for the present in the process of creating a permanent revolution for our times. (McLaren, 2008, p. 480)

My own suggestions for critical classroom practice are outlined in chapter 8.

'Race' and Multicultural Education in the United Kingdom

At the outset, it needs to be stressed that the social, cultural and religious diversity of British society is not a new phenomenon. Britain is a multicultural society and always has been. This is witnessed by the separate existences of England, Scotland and Wales. It is also evidenced by settlement from Ireland and elsewhere in Europe, both in the past and more recently.

Britain's links with Africa and Asia are particularly long-standing. For example, there were Africans in Britain—both slaves and soldiers in the Roman imperial army, the latter occupying the southern part of the British Isles for three and a half centuries (Fryer, 1984, p. 1) before the Anglo-Saxons ('the English') arrived.[8] There has been a long history of contact between Britain and India, with Indian links with Europe going back 10,000 years (Visram, 1986). Africans and Asians have been born in Britain from about the year 1505 (Fryer, 1984; see also Walvin, 1973), and their presence has been notable from that time on.

My concern here, however, is with the educational theory that has developed in the light of the presence in UK schools of the daughters and sons of racialized migrant workers who had entered the United Kingdom from its former colonies as the United Kingdom faced a major labor shortage after World War II.[9]

Multicultural Education and Antiracist Education

Broadly two new types of education were proposed to replace the traditional monocultural approach ('British culture' and 'British values'), which, in reality, was and still is hegemonic. These were multicultural and antiracist

education. Those who advocated the former were predominantly liberals and those who favored antiracist education were mainly Marxists and other Left radicals.

Hazel Carby, 1979 parodied multicultural education in Britain in this brief chronicle:

> *Schools*: We're all equal here.
> *Black students*[10]: We *KNOW* WE are second-class citizens, in housing, employment and education.
> *Schools*: Oh, dear. Negative self-image. We must order books with Blacks in them.
> *Black students*: Can't we talk about Immigration Laws or the National Front?[11]
> *Schools*: No, that's politics. We'll arrange some Asian and West Indian Cultural evenings

As it happens the parody was not that inaccurate. Carby, 1979) goes on:

> It is necessary to ask: who are the socially constituted speakers and initiators of the social practice of the discourse [of multiculturalism]? Clearly, they are not the ethnic minorities themselves but the representatives of dominant social forces to whom 'Blacks' are a problem. Concrete political and economic conditions and contradictions that face both black and white alike are not addressed but are contained within and defected by the concept of multiculturalism.

The multicultural education lobby were given a boost in 1985 with the publication of The Swann Report, *Education for All*. This Report made some of the most wide-ranging suggestions for education in an ethnically diverse society. Amongst these was the suggestion that children in *all* schools should be educated for life in a multicultural society. One of the underlying principles of this suggestion was that if children were taught about each other and each other's cultures, this would help to reduce prejudice, especially amongst white children.

The Swann Report's predominant focus on culture set the trajectory of multicultural education along a superficial line in which children learnt about the food, the clothes and the music of different countries without also understanding the structural and institutional inequalities which had been at the core of community campaigns (Sarup, 1986; Troyna, 1993). The exoticization of minority ethnic group cultures and customs merely served to reinforce the notion that these cultures were indeed 'Other' and drew the boundary more firmly between 'Them', the 'immigrants' or 'foreigners' and 'Us', the 'real' British (Cole and Blair, 2006).

The antiracist critique of monocultural education is that in denying the existence of, or marginalizing the cultures of minority ethnic communities, it was and is profoundly racist. The antiracist critique of multicultural education is that it was and is patronizing and superficial. It was often characterized as the three 'Ss', 'saris, samosas and steel drums' (for a discussion, see Troyna and Carrington, 1990; see also Cole, 1992a). Antiracist education starts from the premise that the society is institutionally racist, and that, in the area of 'race' and culture, the purpose of education is to challenge and undermine that racism. Ten years ago (Cole, 1998b, p. 45) I suggested the way in which an antiracist version of the Australian bi-centennial of 1988 might have been taught in primary (elementary) schools in the United Kingdom (was I writing Chronicles 10 years ago without knowing it?). Here, in order to illustrate the fundamental differences between monocultural, multicultural and antiracist education, I will update this analysis, and extend it to incorporate how traditional monocultural and traditional multicultural approaches might manifest themselves today (these three approaches apply not just to the United Kingdom, but are global in their reach).

In the monocultural classroom children would be taught that Australia was discovered by Captain Cook, an Englishman some two hundred years ago, and that, although Australia is on the other side of the world, the people there are like us, eat the same food and have the same customs and way of life. The climate is much hotter and people can swim on Christmas day, and at many beaches Father Christmas arrives on a surfboard, or even on a surf lifesaving boat. There are still some Aborigines in Australia and the government has recently enacted laws that safeguard their communities against drunkenness and other forms of antisocial behavior.

In the multicultural classroom children would learn that Australia is a multicultural society, just like ours, with lots of different cultures and religions making the country an exciting place in which to live. As well as 'the English', people have emigrated to Australia from most of the rest of Europe, and indeed the world. The multicultural nature of Australian society means lots of different foods, music, dance and national costumes. The Aboriginal people, the original inhabitants, have a thriving culture, and produce very original music and art.

The antiracist classroom would focus on the fact that the indigenous peoples of Australia and their supporters view what happened two centuries ago as an imperialist colonial invasion. Given access to a comprehensive range of resources pertaining to life in Australia, children would discover that in reality multicultural Australia is a racialized capitalist society stratified on lines of ethnicity, class and gender, with Australian-born and English-speaking white male immigrants a the top of the hierarchy and Aboriginal women at the bottom. They would learn about 'land rights' and other struggles, and the economic and ecological arguments pertaining to these rights. They would discover that Aboriginal communities have faced ongoing exploitation and oppression since the invasion, and how this has intensified in recent years. They would relate Australian indigenous struggles against injustice to other struggles for social justice in Australia, and to struggles worldwide.[12]

The Marxist underpinnings of this last approach should be clear in the references to imperialism, colonialism, and the racialized capitalistic nature of Australian society, both historically and contemporaneously.

Up until the late 1990s, with their prognoses that Britain is an institutionally racist society, antiracists were branded by many as 'loony Lefties' and ostracized by the mainstream.[13] In Margaret Thatcher's memoirs (Thatcher, 1993, p. 598), extreme concern is respect about the fact that I was teaching antiracist education in the late 1980s at what was then Brighton Polytechnic. At the time (1987), she also opined, with particular respect to primary math:

> In the inner cities where youngsters must have a decent education if they are to have a better future, that opportunity is all too often snatched from them by hard-left education authorities and extremist teachers. Children who need to be able to count and multiply are learning anti-racist mathematics—whatever that is. (cited in Lavalette et al., 2001)

Similarly, her successor John Major declared at the 1992 Conservative Party annual conference speech:

> I also want a reform of teacher training. Let us return to basic subject teaching, not courses in the theory of education. Primary teachers should teach children to read, not waste their time on the politics of gender, race and class. (cited in ibid.)[14]

It took the Stephen Lawrence Inquiry Report (Macpherson, 1999) to change this. While the Report could have gone further in its castigation of the inherent racism in British society, for antiracists, as I agreed with David Gillborn in chapter 3 of this volume, it is nevertheless a milestone in being the first acknowledgement by the British State of the existence of widespread institutional racism. Leading UK antiracist campaigner and writer, Sivanandan rightly describes the Inquiry as 'not just a result but a learning process for the country at large' (2000, p. 1). He argues that through the course of the Inquiry, 'the gravitational centre of race relations discourse was shifted from individual prejudice and ethnic need to systemic, institutional racial inequality and injustice' (ibid.). The Report led directly to the very progressive Race Relations (Amendment) Act (2000) referred to in chapter 2 of this volume. As I pointed out in chapter 3 of this volume, however, initial official state endorsement of the Stephen Lawrence Inquiry Report's acknowledgement of the existence of institutional racism in the police, the education system and other institutions in UK society was short-lived.

There are further threats to the spirit of the Report, and indeed the promotion even of multiculturalism as a result of The Education and Inspections Act (2006), which came into effect in September 2007. This Act introduced a duty on the governing bodies of maintained schools to promote 'community cohesion', placing a new emphasis on schools to play a key role in building a society with a 'common vision', a 'sense of belonging' and making available similar 'life opportunities' for all. Following Wetherell, Lafleche and Berkeley 2007, Andy Pilkington (2007, p. 14) distinguishes between 'soft' and 'hard' versions of community cohesion. The former views community cohesion as complementing rather than replacing multiculturalism. In addition, the 'soft' version recognizes that the promotion of community cohesion requires inequality and racism to be addressed. The 'hard' version, on the other hand, sees community cohesion and multiculturalism as ineluctably at loggerheads and insists that we abandon the divisiveness evident in multiculturalism and instead should adhere to British values. The hard version of 'community cohesion' is thus monocultural.

When New Labour secretary of state for communities, Ruth Kelly favored the hard version. As Pilkington (2007, p. 14) points out, in announcing the launch of the Commission on Integration and Cohesion in August 2006, Kelly stressed that it was important 'not to be censored by PC',[15] wondered whether multiculturalism was 'encouraging separateness' and emphasized how Britain needed to tackle ethnic tensions (Kelly, 2006, cited in Pilkington, 2007, p. 14). Pilkington (ibid.) notes that her speech glossed over any structural roots of any tensions and stressed to migrants 'their responsibility to integrate and contribute to the local community' (Kelly, 2006, cited in Pilkington, 2007, pp. 14–15). He argues that the tenor of the report can be gleaned from two of its proposals that both warrant separate appendices: the juxtaposition of English and translation services, with a marked preference for the former; and a recommendation that priority should be given in the allocation of funding to groups making links between communities rather than single groups (Commission on Integration & Cohesion, 2007, cited in Pilkington, 2007, p. 15). Pilkington (2007, p. 15) concludes:

> Both these proposals implicitly see multiculturalism and community cohesion as in opposition. They signal a shift away from being accommodating to minority concerns and point to a cohesion agenda where people are required to become less welfare dependent and are instead cajoled into learning English and develop cross-community networks.

In this chapter, I have addressed multicultural education in the United States and the United Kingdom, noting the Marxist-inspired antiracist critiques of the concept on both sides of the Atlantic. In the next chapter, I will consider the arrival of CRT in education in the United Kingdom, concentrating on a Marxist assessment of a 2008 book by one of its leading protagonists, David Gillborn.

Chapter 5

CRT Comes to the United Kingdom: A Critical Analysis of David Gillborn's *Racism and Education*

In this chapter, I address the relatively recent arrival of CRT in educational theorizing in the United Kingdom. In so doing, I focus on the latest book by influential UK 'race' and education theorist David Gillborn in the belief that the growing body of work by Gillborn in the field of CRT is highly likely to consolidate its presence in the United Kingdom. Specifically, I critically discuss Gillborn's views on *Marxists;* on *Marx and slavery;* on *Marx and 'species essence';* on *'White powerholders';* on *racist inequalities in the UK education system;* on *education policy;* on *ability;* on *institutional racism;* on *'model minorities';* on *whiteness and free speech;* on *conspiracy;* and on *'struggling where we are'* against *'the powers that be'.*

With respect to multiculturalism and racism, it is true to say that antiracist education, informed by Marxism and other forms of radical thinking, was the predominant Left theoretical position in the United Kingdom right up to the twenty-first century.[1] On November 20, 2006, however, there came a challenge from across the Atlantic. The first ever international Critical Race Theory (CRT) seminar in the United Kingdom took place at the Education and Social Research Institute at Manchester Metropolitan University. CRT is very new to the United Kingdom, and has, as yet, few adherents here. These few are mainly working in the field of education. Indeed, CRT's main UK-based protagonists, David Gillborn (e.g., Gillborn, 2005, 2006a, 2006b, 2008), John Preston (e.g., Preston, 2007) and Namita Chakrabarty (e.g., Chakrabarty, 2006a, 2006b; Charkabarty and Preston, 2006, 2007), all educationists, presented papers at the conference. At the British Education Research Association Annual Conference in September, 2007, at least six papers were CRT-focused, including a symposium, entitled, 'Guess who's coming to Bera? Has critical race theory arrived in UK education research?' (BERA, 2007). At the 2008 BERA Conference papers on CRT were delivered.[2]

Elsewhere (Cole and Maisuria, 2007) we have referred to David Gillborn as 'arguably the most influential "race" theorist within education in Britain'. Gillborn's authority as a leading UK theorist of CRT and Education is likely to be enhanced with the publication of *Racism and Education: coincidence or conspiracy* (Gillborn, 2008). The blurb on the back cover of Gillborn (2008) describes the book as 'the first major study of the English education system using "critical race theory"'. The rest of this chapter will be devoted to a discussion of this book (I have also referred to parts of Gillborn's text throughout this volume). I would like to say at the outset, that I am in full agreement with its main conclusion that 'education policy is not designed to eliminate race inequality but to sustain it at manageable levels' (back cover of Gillborn (2008)).

That the book is written uncompromisingly within the CRT framework is made clear in the first paragraph of Chapter 1 (the Introduction to the book), when Gillborn (2008, p. 1) states that 'race inequality should be placed centre-stage as a fundamental axis of oppression'. Following some remarks, uncontroversial from an *antiracist* perspective, about 'race' (there is 'no such thing') (p. 2) and racism (a wide-ranging definition is needed) (pp. 3–4),[3] Gillborn (pp. 5–19) presents the CRT Chronicle referred to in the appendix to chapter 3 of this volume. As noted in that appendix, the Chronicle features a conversation between an older white professor and a young black male student. The two take part in an engaging conversation which, as they move from lecture room to pub to fish and chip shop to the Professor's office, addresses the subtitle of Gillborn's book—whether 'race' inequality is a coincidence or a conspiracy (the conversation is resumed in the Conclusion to Gillborn's book).

On Marxists

In true postmodern style the Chronicle also allows the Professor (partly Gillborn, but not totally—p. 5)[4] to introduce CRT, and outline the chapters of Gillborn's book. It is in the section entitled *Critical Race Theory*, which is the Professor's/Gillborn's introduction to CRT, that I find my first major disagreement with the Professor/Gillborn (p. 13). The Professor states 'I don't think there's anything in CRT that a serious antiracist would have a problem with' (p. 13). My disagreement stems from the fact that I very much consider myself a serious antiracist, and from a number of conversations and correspondences over the years with Dave Gillborn, I know that he would concur with this. He would also acknowledge that I have been engaged in writing Marxist critiques of CRT, of which this present volume forms part, and therefore that there most definitely *is* a lot in CRT that certain antiracists would have a number of problems with. Indeed in the second chapter of his book, Gillborn (p. 20) makes reference to such a critique, a paper by Alpesh Maisuria and myself (Cole and Maisuria, 2007). Unfortunately in stating that in this paper, our position is that CRT 'gives undue attention to racism rather than class divisions', he greatly oversimplifies our argument. What we actually set out to do, in similar fashion to my arguments in chapter 2 of this volume, is to make the case, *in order to facilitate a serious and in-depth understanding of racism*, that CRT, in its advocacy of 'white supremacy', and in its *pre-eminence* of 'race' over class (Cole and Maisuria, 2007), is not able to attain such an understanding. As in chapter 2 of this volume, in Cole and Maisuria (2007) we commend the Marxist concepts of racialization and xeno-racialization as having the best purchase in explaining manifestations of racism, Islamophobia and xeno-racism in contemporary Britain.

Gillborn (p. 37) further misrepresents our position, when, in referring to Cole and Maisuria (2007) he states that 'a conceptual debate with Marxist orthodoxy may simply be redundant because by definition Marxists place class in a position that supersedes all other forms of exclusion'. There are two responses I would like to make to this assertion. The first is that Gillborn knows that Maisuria and I are keen to debate racism with critical race theorists.[5] I believe that a conceptual debate between Marxism and critical race theory is very important. Indeed, I have engaged in such a debate with Gillborn and other Critical Race Theorists, both face-to-face, and in written form (e.g., Cole, 2008e, 2009a, 2009d; Mills, forthcoming, 2009) for several years. Moreover, as should be clear by now, this is one of the major purposes of this volume: to engage in a conceptual debate with Critical Race Theorists. My second response is that Gillborn is also fully aware that an analysis of racism from a Marxist perspective, rather than an analysis of class, has been one of the central features of my writing over a period of over two decades (I recently (Cole, 2007c, p. 14) described racism as 'one of the key issues facing the world in the twenty-first century'), and thus there are ample opportunities for him to debate with me my Marxist analyses of 'race' and racism, and, of course, the analyses of other writers.

As if to further alienate Marxists, Gillborn (pp. 37–38) goes on to approvingly quote Ricky Lee Allen (2006) who, according to Gillborn (2008, p. 37), 'views contemporary academic Marxism as an exercise of White power'. Arguing stridently against any alliance with Marxists, Allen describes the ascendancy of CRT as a historic rift and a 'much needed shift' (2006, p. 9, cited in Gillborn, 2008, p. 37).[6] It is disappointing that Gillborn seems to want to foreclose discussion with Marxists.

On Marx and Slavery

Gillborn's hostility to Marx is underlined when he refers to some of Mills' work on the relationship between 'White Marxism and Black Radicalism'. He cites Mills (2003, p. xvii) as claiming:

> critical race theory is far from being an adjunct to, or outgrowth of, critical class theory; in fact, it long predates it, at least in its modern Marxist form. Long before Marx was born, Africans forcibly transported as slaves to the New World were struggling desperately to understand their situation; they were raising the issues of social critique and transformation as radically as—indeed

even more radically than—the white European working class, who were after
all beneficiaries of and accessories to the same system oppressing blacks. (cited
in Gillborn, 2008, p. 38)

Gillborn's (2008, p. 38) comment is that 'Mills' point is extremely powerful'.
Gillborn goes on point out that Marx moved to London in 1849, more than
a decade before slavery was abolished in U.S. territories (ibid.). 'These simple
facts', Gillborn states, 'make the minimal presence of race in Marx's analyses
all the more damning' (ibid.).

It is difficult to understand what both Mills and Gillborn are implying. I
will deal with Mills' quote and Gillborn's comments on the quote in turn.
With respect to the quote, Mills seems to be suggesting five things: (1) that
the struggle against racism predates the *modern* European class struggle;
(2) that slaves' analyses and struggles were an early form of critical race
theory; (3) that slaves were more radical that the white European working class;
(4) that the white working class were beneficiaries of slavery; and (5) that
they were accessories to it. With respect to (1), this seems to be truism. As
far as (2) is concerned, given that critical race theory grew out of critical
legal studies in the 1980s, a fact heralded by those central to the movement
(see chapter 2 of this volume), it is difficult to make sense of Mills' assertion.
That slaves were more radical than the white working class (3) is difficult
to quantify. It really depends what Mills means by 'radical'. With respect to
(4), that the white working class were beneficiaries to slavery, this is true in
the sense that they accrued some benefits from capitalist plunder. Finally,
whether the white working class were accessories (5) needs to be seen in the
context of the success of the interpellation process (interpellation is discussed
in chapter 1 of this volume). To merely list the class as 'accessories' implies
conscious rational choice outside the confines of ideological processes.

If my response to these five points makes any sense, it is difficult to
understand why Gillborn finds them 'extremely powerful'. As to his development of Mills' assertions, while I accept Gillborn's point that there is a
minimal presence of 'race' in Marx's writing, Gillborn seems to be implying
that, given that slavery existed in the U.S. territories when Marx arrived in
London, that Marx should have written about it, but did not, and should
therefore be 'damned' for it. In actual fact, Marx, a leading European abolitionist, was London Correspondent for the radical anti-slavery 'New York
Daily Tribune' (Laskey, 2003, p. 1). During the U.S. Civil War, Marx urged
and organized English textile workers to support the blockade against the
Confederacy, even though it was not in their immediate economic interests
and also led to massive layoffs as a result of the cut off of imported cotton
(Marx, 1862, p. 153). Writing about the importance of working class extra-
parliamentary activity, Marx described working class disgust and action
against the Confederacy as 'admirable', 'incredible', and as 'more striking'
than other demonstrations (e.g., against the Corn Laws and the Ten Hours
Bill) because of its unambiguous spontaneity and persistence (ibid.). Marx
saw the action as 'new, brilliant proof of the indestructible staunchness

of the English popular masses' (ibid.), and reported with great enthusiasm on 'a great *workers' meeting* in Marylebone, the most populous district of London' (ibid.) which served 'to characterise the "policy" of the working class' (ibid.). At that meeting, the following motion was passed unanimously:

> This meeting resolves that the agents of the rebels...are absolutely unworthy of the moral sympathies of the working class of this country, since they are slaveholders as well as the confessed agents of the tyrannical faction that is at this very moment in rebellion against the American republic and the sworn enemy of the social and political rights of the working class in all countries. (cited in Marx, 1862, p. 153)

At the same meeting another motion, expressing 'the warmest sympathy with the strivings of the Abolitionists for a final solution to the slave question' (cited in Marx, ibid.) was also adopted unanimously. The final motion, also unanimous, was to forward to the U.S. government a copy of the resolutions 'as an expression of the feelings and opinions of the working class of England' (ibid.).

Marx took this strong Abolitionist position because, as he wrote to Engels on the eve of the Civil War, the uprisings of slaves in the United States and of serfs in Russia were the 'two most important events' taking place in the world (Marx 1860, p. 5). Marx expresses his views on slavery succinctly in *Capital Volume 1*:

> In the United States of North America, every independent movement of the workers was paralysed as long as slavery disfigured part of the Republic. Labour cannot emancipate itself in the white skin where in the black it is branded. (Marx (1887) [1965], p. 301)[7]

On Marx and 'Species Essence'

On page 41 of chapter 3, Gillborn (2008) states that CRT argues 'strongly against any comforting belief in the essential goodness of the human spirit'. In chapter 3 of this volume, I expressed agreement with Gillborn, in his conclusion that The Lawrence Inquiry was not agreed to by a benign state that wanted to put right an injustice, but was rather in the wake of protests and demonstrations. However, given that the context on Gillborn's page 41 is a discussion about whether racism is permanent, and given that for Gillborn (2008, p. 41) this is 'a moot point', more seems to be being said here. Gillborn seems to be making a more general, more ahistorical[8] point about humankind. Marx would not agree, since he related our humanity to the capitalist mode of production, which he believed stifles the worker's 'species essence'. In order to understand what Marx meant, it is necessary to briefly consider Marx's theory of alienation. Marx attributes four types of alienation, a fundamental condition of labor under capitalism, which prevented humankind from realizing its species-being and establishing an objectively better socialist society. These are described by Gordon Marshall (1998) as follows:

> alienation of the worker from his or her 'species essence' as a human being rather than an animal; alienation between workers, since capitalism reduces labour to a commodity to be traded on the market, rather than a social relationship; alienation of the worker from the product, since this is appropriated by the capitalist class, and so escapes the worker's control; and, finally, alienation from the act of production itself, such that work comes to be a meaningless activity, offering little or no intrinsic satisfactions.

Marshall (ibid.) goes on to argue that the last of these 'generates...feelings of powerlessness, isolation, and discontent at work—especially when this takes place within the context of large, impersonal, bureaucratic social organizations'. In Marx's own words, this is how the alienation of labor affects the worker:

> [It] mortifies his flesh and ruins his mind. Hence, the worker feels himself only when he is not working; when he is working, he does not feel himself. He is at home when he is not working, and not at home when he is working. His labour is, therefore, not voluntary but forced, it is *forced labour*. It is, therefore, not the satisfaction of a need but a mere *means* to satisfy needs outside itself. Its alien character is clearly demonstrated by the fact that as soon as no physical or other compulsion exists, it is shunned like the plague.[9]

Thus, workers under capitalism cannot come to full self-realization. To be alienated is to be separated from one's essential humanity. We can only fulfill our species essence through freely chosen labor, in a collective and cooperative society. Only in such a society can our 'essential goodness', to use Gillborn's terminology, come to fruition.

On 'White powerholders'

Gillborn's ahistorical stance is also revealed in the final part of chapter 2 when he refers to Derrick Bell's *Chronicle of the Space Traders* (Bell, 1992b).[10] As Gillborn (2008, pp. 41–42) explains, this is a short story that recounts what happens when aliens visit the United States and offer a simple trade: as much gold and technology as is needed to solve the U.S. economic and environmental crises, in return for every African American being taken away to an unknown fate. There follows a flurry of activities including secret meetings between politicians and capitalists, where the capitalists express concern at losing an important market, and a useful scapegoat. In the end, millions of African Americans enter the trader ships in chains just as their forebears entered the New World in the first place. Bell (1992b, p. 13) is right to conclude that 'an ultimate sacrifice of black rights—or lives' is '[e]verpresent, always lurking in the shadow of current events' (cited in Gillborn, 2008, p. 42), and Gillborn (2008, p. 42) is right that Jean Charles de Menezes, an innocent man killed by police with seven shots to the head, was such a sacrifice. However, this is not the decision of 'whites' (Bell, 1992b, p. 13, cited in Gillborn, 2008, p. 42), or of 'White powerholders' (Gillborn, 2008, p. 42), *viewed as a perpetual omnipresent and ubiquitous ahistorical and essentialist force,* as Critical Race Theorists would have us believe.

These white power holders need to be situated economically, politically and ideologically. They are, as Bell rightly notes in the case of his Chronicle, 'politicians' and 'capitalists'. With respect to de Menezes, they are agents of the Repressive State Apparatus, acting under orders in the wake of mounting Islamophobia (see chapter 2 of this volume), in a period two weeks after the events of July 7, 2005 (7/7). These bomb attacks in London need themselves to be sent in the context of UK-supported U.S. imperialist adventures in Iraq and Afghanistan (see chapter 6 of this volume).

Gillborn (2008, pp. 42–43) describes a radio program, almost a year after 7/7, which focused on a high profile raid by 250 police in East London which involved the shooting at close range (not fatal) of a Muslim man. Two Muslim men were taken into custody, and released later without charge.

In response to the question, 'are you prepared for the police to make mistakes sometimes?', the presenter of the UK radio station, *Five Live* read the following sentiments from listeners without comment:

'[i]f there were no Muslim terrorists, there would be no police raids on Muslims. It's simple'
'It's *good* that these Muslims, Arabs and Asians are having it rough here' and 'I'd rather the odd one got shot than a relative of mine got blown up'. (cited in Gillborn, 2008, pp. 42–43)

This is not as Gillborn (2008, p. 42) maintains 'one final piece of evidence on the matter of the Space Traders'; it cannot be explained in Bell's (1992b, p. 13, cited in Gillborn, 2008, p. 42) words simply as a consensus among whites 'that a major benefit to the nation justifies an ultimate sacrifice of black rights—or lives', nor by a 'lack of essential goodness in the human (white) spirit'. Rather this needs to be situated in the context of British imperialism and its aftermath, as well as New U.S. imperialism and the global frenzy for more and more surplus value (see chapter 6 of this volume). The listeners' comments cannot be separated out from successful interpellation, from the successes of Ideological Apparatuses of State (Althusser, 1971), in particular in this case, the media, and perhaps particularly *The Sun* or *The Daily Mail*. It is important for Bell to mention the discussion between politicians and capitalists in his *Chronicle of the Space Traders*, but these reactions also need grounding in the real world. A white supremacist fantasy story may be useful for its shock value (see chapter 3) but it remains that—'a white supremacist fantasy story' unless it is situated historically, and in the context of specific eras in the capitalist mode of production, and related to old or new imperialisms.

I would like to underline my point by referring to a statement which appeared, at the time of writing in the *Guardian* newspaper: 'It is unfortunate that people got killed' (cited in McGreal, 2008b). The young man reported in the *Guardian* goes on to say, '[b]ut they had to go. They do not belong here taking jobs'. 'Let them go back to Zimbabwe and solve their own problems instead of bringing them here. We have enough problems of our own' (cited in ibid.). However, this is a long way from the United Kingdom, and the context very different and totally unrelated to any notion of 'White powerholders' or 'white supremacy'. This chilling vignette could be almost anywhere in the world where migrants are on the receiving end of neoliberal capitalism. In fact it is in South Africa, and the young man is a black South African. As the *Guardian* writer, Chris McGreal (2008b) explains

> No one in Cleveland squatter camp seemed to know the names of the five burned or bludgeoned bodies. They were referred to simply as Zimbabweans, though no one could even be sure they were that. It was enough that they were foreigners accused of taking jobs, houses and women—or of leading a crime wave—by the mobs that killed them and drove hundreds of others from their homes. About 50 people were taken to hospital with gunshot and stab wounds as the gangs smashed their way in to the dozen or so foreign-owned shops in Cleveland, in the south of Johannesburg.

McGreal explains that '[h]ostility to Africans from other parts of the continent has long been rife in South Africa but has escalated with the arrival of the Zimbabweans'. They are 'popular with local employers because many are well educated, speak good English and are seen as working harder than South Africans'. McGreal concludes that seven people were killed earlier in the year, including a Somali, Zimbabweans and Pakistanis and two Somali shop owners. Zimbabwean Grace Muzenda tells McGreal, 'they always hated us. We thought this might happen' (McGreal, 2008b). The racism reported here, fostered by capitalists' universal desire for the generation of extra surplus value, is not color-coded. As with the reactions of the listeners to *Five Live* these events in South Africa need to be understood in context. Since the post-apartheid governments have gone down the road of neoliberal capitalism, rather than social democracy let alone socialism, improvements in housing, electricity, and water supply, health, and education that were anticipated after the end of apartheid have not materialized (Talbot, 2008). Most of the poor live in conditions that are as bad as or worse than under apartheid (ibid.). At the same time, a tiny minority of the elite in the African National Congress have become fabulously wealthy. The gap between the rich and poor has widened under the ANC government. Black empowerment has created a layer of rich businessmen (ibid.). Echoing developments elsewhere, 'Government ministers have consistently demonized illegal immigrants, while at the same time making it extremely difficult for them to gain legal status' (ibid.).

The longer term historical context in this case consists of a complex history of three centuries of capitalism and imperialisms, starting with the Dutch and British colonization of southern Africa, and the intense form of capital accumulation that was apartheid, as well, perhaps, as the categorization of people into 'tribes' by nineteenth century and early twentieth century colonial anthropologists and others (Vail (ed.) 1989). Rule in these three centuries up to and including apartheid most certainly could be described as the hegemony of 'white powerholders', but not any more.

On Racist Inequalities in the UK Education System

In chapter 3 of Gillborn (2008), Gillborn maps, critically analyses and meticulously explains racist inequalities in the UK education system. After twenty-five pages of painstaking analysis, he comes to the conclusion that while a complex picture emerges, 'particular minoritized groups (Black students and their peers of Pakistani and Bangladeshi heritage) continue to be significantly less likely to achieve the key benchmarks when compared to White peers of the same gender' (Gillborn, 2008, p. 69). Moreover, white students are the *only* group to show an increase in the number of higher grade GCSEs in every survey since the late 1980s, and young people in each of the 'black' monitoring categories are more likely to be permanently excluded from school (ibid.). Gillborn (2008, pp. 63–64) introduces the CRT concept of 'locked-in inequality' to describe 'an inequality so deep rooted and so large that, under current circumstances, it is a practically inevitable feature of the education system'. Chapter 3 also contains a useful discussion of 'Gap Talk', whereby '[t]alk of "closing" and/or "narrowing" gaps operates as a discursive strategy whereby statistical data are deployed to construct the view that things are improving and the system is moving in the right direction' (Gillborn, 2008, p. 65). As Gillborn puts it, this is more than mere *reporting*: rather statistics are conveyed with a particular tone and emphasis that encourages positive interpretation (ibid.). Such revealing insights are of use and interest to all of us involved in the antiracist struggle and concerned with how the racist capitalist state tries to present itself in a non-racist light, but they do not incline me any more towards CRT as an alternative to Marxism. Rather they are indicative of the continuing racialization of Asian and black students which has its origins in the British Empire (see chapter 2 of this volume).

On Education Policy

Chapter 4 deals with 'race' education policy under New Labour. Gillborn (2008, p. 75) describes the first four years as 'naïve multiculturalism' because although there was evidence of a limited commitment to equity, apart from the decision to set up separate faith schools and the Stephen Lawrence Inquiry Report (Macpherson, 1999), this was largely superficial, consisting of 'rhetorical flourishes that left mainstream policy untouched' (p. 75).

Gillborn (2008, pp. 76–80) refers to the second period, between 9/11 and 7/7 (2001–2005), as 'cynical multiculturalism'. He describes it (pp. 76–77) as a period when the government continued

its *rhetorical* commitment to ethnic diversity and race equality...[in] a cynical attempt to retain the appearance of enlightened race politics while simultaneously pursuing a policy agenda that increasingly resembled the earlier assimilationist/integrationist phases...where the voices and concerns of White people were openly accorded a position of dominance.

In the context of 9/11, Gillborn (2008, p. 77) cites Adams and Burke (2006, p. 991) who state that the racialization of that event and the resulting demonization of entire minoritized communities simply could not have happened had the attackers been part of the white racial majority. While this may be true for the white 'racial' *majority,* one has to wonder whether Adams and Burke (Gillborn does not tell us if they are Critical Race Theorists) and Gillborn, as a Critical Race Theorist, are forgetting the fact that the racialization of the Irish, a white 'racial *minority'* in Britain, intensified at times of IRA activity, particularly when actually located in Britain (Walter, 1999, p. 319).

The third period, from 7/7 (2005) up to the present is described as 'aggressive majoritinarianism' (Gillborn, 2008, pp. 81–89) (this would equate with Wetherell, Lafleche, and Berkeley's, 2007, concept of 'hard community cohesion' discussed earlier in this chapter). On July 7, 2005, as noted earlier in this volume, a coordinated series of explosions in London killed fifty-two people, and injured a further seven hundred. Gillborn (2008, p. 81) argues that the attacks heightened 'still further the retaliatory confidence of politicians and the media', and that their mood from one of retaliation to 'aggressive majoritarianism' 'where Whites now took the initiative in promoting ever more disciplinary agendas' (ibid.). As he concludes (ibid.):

> The rights and perspectives of the White majority were now asserted, sometimes in the name of 'integration' and 'cohesion' (the code words for contemporary assimilationism) but also simply on the basis that the majority disliked certain things (such as Muslim veils) and now felt able to enforce those prejudices in the name of common sense, integration and even security.

For Marxists, any discourse is a product of the society in which it is formulated. In other words, 'our thoughts are the reflection of political, social and economic conflicts and racist discourses are no exception' (Camara, 2002, p. 88). While such reflections can, of course, be refracted and disarticulated, dominant discourses (e.g., those of the Government, of Big Business, of large sections of the media, of the hierarchy of some trade unions) tend to directly reflect the interests of the ruling class, rather than 'the general public'. The way in which racialization connects with popular consciousness, however, is via 'common sense'. 'Common sense' is generally used to denote a down-to-earth 'good sense' and is thought to represent the distilled truths of centuries of practical experience, so that to say that an idea or practice is 'only common sense' is to claim precedence over the arguments of Left intellectuals and, in effect, to foreclose discussion (Lawrence, 1982, p. 48). As Diana Coben (2002, p. 285) has noted, Gramsci's distinction between good sense and common sense 'has been revealed as multifaceted and complex'. For common sense:

> is not a single unique conception, identical in time and space. It is the 'folk-lore' of philosophy, and, like folklore, it takes countless different forms. Its most fundamental characteristic is that it is...fragmentary, incoherent and inconsequential. (Gramsci, 1978, p. 419)

A clear example of aggressive majoritarianism is Tony Blair's assertion that '[o]ur tolerance is part of what makes Britain, Britain. So conform to it; or don't come' (cited in Gillborn, 2008, p. 83). As Gillborn points out, citing Karen Chouhan (2006), Britain's tolerance is based on intolerance' (Gillborn, 2008, p. 83). Gillborn (ibid.) also notes a marriage of 'aggressive majoritarianism' and 'retaliatory confidence' when Blair threatened in the same speech: 'we're not going to be taken for a ride' (cited on p. 84). More aggressive majoritarianism followed—presented not as a restriction but as part of equal opportunities—when Blair stated that as a requirement of 'equal opportunity', 'cohesion' and 'justice', the use of English should be a condition for citizenship (cited on pp. 84–85), while Gordon Brown (then Chancellor of the Exchequer), not to be out-done, announced that 'community work' should also be part of become a British citizen (cited on p. 85). As Habib Rahman, chief executive of the Joint Council for the Welfare of Immigrants (cited in ibid.), has pointed out '[c]ompulsory community service is usually imposed as a non-custodial penalty for a criminal offence'.

In his Conclusion to the chapter, Gillborn (2008, p. 86) suggests that contemporary 'race' equality 'in key aspects...is as bad, and in some cases worse, than anything that has gone before'. Whether this is the case or not, the point is to explain *why*. 'It is hardly surprising', Gillborn (ibid.) notes 'that the state should prioritize its own survival and, more specifically, that ruling parties should seek to maintain popularity with a majority of the electorate'. For Gillborn, the state is a white supremacist one, whereas for Marxists it is a neoliberal pro-imperialist capitalist state, and it is this fact that explains its policies and actions (see chapter 6 for a discussion of neoliberalism, global capitalism and imperialism).

Tracing social policy in relation to 'race' in the post-war period, Gillborn (2008, p. 88) sees the 1950s as indicative of early 'assertions of the supremacy of the "host" society', where white parents needed assurance that 'minoritized students were not damaging *their* children's education'. Whereas for Gillborn this is related to 'white supremacy', from a Marxist perspective such racism needs to be understood in relation to the differential racialization of the subjects of the former imperial subjects, whose children were now entering the education system. An important element of this is the continuity in the differential portrayal of Asian and black children respectively (see Cole and Blair, 2006).

Gillborn (2008, p. 88) goes on suggest that from the 1950s, through Thatcherite 'new racism' to 'New Labour's aggressive majoritarian "common sense" assimilitationism', the constant assumption has been that 'the interests, feelings and fears of White people must always be kept centre stage'. Here, it needs to be pointed out that these 'interests, feelings and fears' are not kept centre stage for the *benefit* of the white working class. Rather that class is interpellated as sharing common interests with the capitalist state, when in fact its interests are diametrically opposed (see pp. 159–160 of this volume). From the Empire Windrush[11] to the entry of Poland into the European Union, it has been useful for the British capitalist state to have a ready supply of cheap labor, whose presence is publicly vilified in order for the state to maintain hegemony over the longer-residing population.

Gareth Dale (1999, p. 308) explains the contradiction between capital's need for (cheap) flexible labor and the need to control the workforce by racializing potential foreign workers:

> On the one hand, intensified competition spurs employers' requirements for enhanced labour market flexibility—for which immigrant labour is ideal. On the other, in such periods questions of social control tend to become more pressing. Governments strive to uphold the ideology of 'social contract' even as its content is eroded through unemployment and austerity. The logic, commonly, is for less political capital to be derived from the [social contract's] content, while greater emphasis is placed upon its exclusivity, on demarcation from those who enter from or lie outside—immigrants and foreigners.

Gillborn (2008, p. 88) concludes chapter 4 by referring to another 'common element in British social and educational policy', which he describes as 'the strategic deployment of White racial violence as a limit to policy and a threat against those who would challenge the chosen orthodoxy'. He has in mind the late 1950s when white mobs, partly organized by fascists, terrorized Asian and black people on the streets of London and Nottingham (ibid.). He cites subsequent prime ministers as using this precedent as 'a more or less overt threat' of further white violence. Once again, for Marxists, the problem is not one of 'white violence'. It is more complicated than this. From a Marxist perspective, violence against racialized groups *must* be seen in the context of the respective requirements of capitalists and their allies in specific historical periods, the success with which workers are interpellated by the requirements of the capitalist state, and by the historical and current realities of old and new imperialisms (see chapters 2 and 6).

In chapter 3 of this volume, I have already noted that, reporting on national assessment mechanisms, Gillborn (2006b) argues convincingly that these actually *produce* inequality for black pupils, with the caveat that such facts need relating to racialized capitalism. This line of argument is developed in chapter 5 of Gillborn, 2008. Gillborn states that the data that he presents suggests that 'the "assessment game" is rigged to such an extent that if Black children succeed as a group, despite the odds being stacked against them, it is likely that the rules will be changed to re-engineer failure' (Gillborn, 2008, p. 91). He argues that '[d]espite the rhetoric of "higher standards for all"' many black students are locked into a system with fixed grade limits which literally prevents them from getting the highest grades, and describes how a new system of assessment for five year old children wipes out the only part of the whole system where black children were successful (p. 116). Finally he notes that black/white inequality for five-year-olds 'is growing at the same time that teachers' training in the new system is supposedly reaching new heights' (ibid.). Gillborn (p. 117) concludes the chapter by arguing that the evidence suggests that not only does assessment *produce* inequality, it sustains it as well.

As before, I would indicate full agreement with this insight of the workings of the racist capitalist state in Britain, but also as before, it does not incline me any further towards Critical Race Theory. Moreover, none of this detracts from the importance of the effect of social class. Indeed, this was noted by Gillborn himself writing with Heidi Mirza in his pre-CRT days (see Gillborn and Mirza, 2000). (For analyses of the crucial determinant of social class, see Hill, 2008a, 2009c, who refers regarding statistical and analytical data, to Dehal, 2006; Abbass, 2007; Demie and Tong, 2007; Demie et al., 2007; Strand, 2007; and for conceptual analysis see Kelsh and Hill, 2006). While Gillborn (2008 p. 69) does state that 'the data certainly confirm that social class background is associated with gross inequalities of achievement at the extremes of the class spectrum' (and later in the book notes 'the strong association between social class and educational achievement' [Gillborn, 2008, p. 147], and that Chinese and Indian students are 'significantly less likely to experience economic disadvantage and more likely to attend private schools' [p. 160]), he states that 'class does not appear to be significantly significant for all groups' (2008, p. 69.). He adds, in order to retain his post-2000 faith in CRT, '[T]he growing emphasis on students in receipt of free school meals (FSM)...projects a view of failing Whites that ignores the five out of every six students who do not receive FSM' (Gillborn, 2008, p. 69). As Hill (2008a) puts it, while Gillborn, 2008 gives specific recognition that social class is 'raced' and gendered, he 'gives...very substantially less...recognition that 'race' is classed (and gendered)'. Hill (2008a) argues that the UK data 'does not show an overall pattern of White supremacy' and concludes that with respect to Gillborn's 2008 book:

> While his work is not silent on social class disadvantage and social class based oppression, his treatment of social class analysis is dismissive and his treatment of social class underachievement education and society, extraordinarily subdued. (Hill, 2008b; see also Hill, 2008a, 2009c)

Figures released at the time of writing (December, 2008) reveal that 'a small group of traveller pupils' (Curtis, 2008) are now the lowest achievers with respect to GCSE results, followed by 'white British' working class boys (ibid.).

On Ability

In chapter 5, Gillborn (2008, pp. 110–116) also deals with the concept of 'ability', pointing out that much of the 2005 Government publication,

'Higher Standards, Better Schools for All' (Department for Education, Schools and Families, 2005) embodies the kind of assumptions about 'ability' that were around more than sixty years ago, and which have repeatedly been debunked (Gillborn, 2008, p. 110). Here I am in total agreement with Gillborn. Indeed, in chapter 4 of this volume, I made brief reference to the *Learning Without Limits* (*LWL*) project of Hart et al. (2004) that critiques the notion of 'fixed ability', and to my Marxist assessment thereof (Cole, 2008b). As I put it in Cole, 2008c, p. 454:

> Fixed Ability adversely affects the working class as a whole. This is because while that which they are perceived as 'able' to do needs to constantly adapt to changes in the capitalist mode of production, the working class are always labelled as having 'limited potential' in order that their subordinate structural location in a hierarchical capitalist society appears as 'natural' rather than as driven by capitalist imperatives. This appearance of 'naturalness' has to be constantly reinforced ideologically.

In England and Wales, school tests (children sit more exams that anywhere else in Europe) play a key role in this ideological process (D. Davis, 2008, p. 12). Dave Davis (ibid.) encapsulates how it works:

> would you rather work with someone who when faced with a problem removes themselves into a quiet room and ponders it in isolation for one hour and forty minutes? Or someone who attempts to find the answer by asking other people and looking things up on the internet—in other words, cheating. In fact, all exams do is test the ability of children to sit tests.

Davis concludes that the tests perform one crucial function—'they help maintain the myth that we live in a meritocracy'. They justify superior salaries by 'superior ability' (ibid.). Speaking against and working against fixed ability and differentiation is essential for all those who wish to militate against the stifling of the capacity and unlimited potential of the working class. The *Learning without Limits* (*LWL*) project is one such attempt to do this.

Gillborn (2008, pp. 114–116) charts the way in which New Labour's 'gifted and talented' scheme, claimed by the Department for Education and Employment that it will operate 'regardless of ethnic background' (cited in Gillborn, 2008, p. 114) whereas in fact recent inspection reports show 'only small numbers from Pakistani, Black African and Black Caribbean heritages' appeared on NAGTY (National Academy for Gifted and Talented Youth) courses (Gillborn, 2008, p. 116).

I am in full agreement with Gillborn's (2008, p. 117) overall conclusion to chapter 5 which is:

> Until we address the presence of racism, as a fundamental defining characteristic of the education system, the present situation is unlikely to change in any meaningful sense, irrespective of superficial rhetorical commitments to inclusion, civil rights and social justice.

On Institutional Racism

Chapter 6 of Gillborn (2008) focuses on the Stephen Lawrence case, and I have commented on Gillborn's analysis of this case in chapter 3 of this volume. I would just like here to make a comment on the definition of 'institutional racism' in the Stephen Lawrence Inquiry Report (Macpherson, 1999), which Gillborn (2008, p. 152) suggests 'clearly attempts to recognize the complex (sometimes hidden) nature of racism'. Institutional racism is defined in the Report as:

> The collective failure of an organisation to provide an appropriate and professional service to people because of their colour, culture, or ethnic origin. It can be seen or detected in processes, attitudes and behaviour which amount to discrimination through unwitting prejudice, ignorance, thoughtlessness and racist stereotyping which disadvantage minority ethnic people. (Macpherson, 1999, 6.34)

This definition was given a formal seal of approval by its having been read in the House of Commons on 24 February, 1999, by the then Home Secretary, Jack Straw. It is interesting to note, however, that in repeating the definition verbatim in his speech to the House, Straw stresses the word, 'unwitting' (http://news.bbc.co.uk/1/hi/uk/285553.stm—audio link available).

While the official acknowledgement of the existence of institutional racism is a most welcome development, and is a considerable conceptual leap from notions of personal prejudice, I disagree with Gillborn that Macpherson's definition is complex. I believe that there is a need to situate the concept, historically, economically and politically. I have argued at length elsewhere that racism in British society may be viewed as a continuous process from the origins of the Welfare State up to the present, both in general terms (Cole, 1992b; Cole and Virdee, 2006) and with particular respect to education (Cole, 1992b; Cole and Blair, 2006), and that racism, institutional or otherwise, cannot be understood without situating it within historic, economic and political processes. There is a need, therefore, to incorporate these dimensions in a definition of institutional racism. The Marxist concept of racialization needs also to be included to move away from the nebulous and ahistorical definition of institutional racism provided by Macpherson. I believe such a definition needs also to include 'common sense', which I argued earlier in this chapter, connects racialization with popular consciousness. Finally, in line with my definition of racism in chapter 2, I would also want to add *intentional* as well as unintentional or unwitting racism. Institutional racism is thus reformulated as:

> Collective acts and/or procedures in an institution or institutions (locally, nation-wide, continent-wide or globally) that intentionally or unintentionally have the effect of racializing, via 'common sense', certain populations or groups of people. This racialization process cannot be understood without reference to

economic and political factors related to developments and changes in national, continent-wide and global capitalism.[12]

On 'Model Minorities'

In chapter 7, Gillborn (2008; see also chapter 1 of this volume) looks at the issue of 'model minorities'. Gillborn (2008, p. 151) warns of 'the dangers of teachers' assumptions about innate racialized "potential"', for example, the 'exceptionally high expectations that many teachers hold about Indian and Chinese students [that] are the flip side of the same coin that involves the demonization of Black students' (p. 153). It is important to point out that whatever the academic success rates of specific ethnic groups, there is still the reality of racist harassment and labor market exclusion (Gillborn, 2008, p. 157). Gillborn (2008, pp. 156–157) suggests that 'White people draw considerable benefit from the existence of so-called model minorities: the stereotype provides a strong rhetorical counter to accusations of racism and unfairness'. It is difficult to see how white people in general derive such benefit, but easy to see how the racist capitalist state and its supporters do.

On Whiteness and Free Speech

The subject matter of chapter 8 is 'whiteness' with Gilborn (2008, p. 162) asserting that white *privilege* is 'too soft a word; this is about *supremacy* (original emphasis)'. I have critiqued the notion of 'white supremacy' at length in chapter 2 of this volume, so will limit myself here to a couple of observations. Gillborn's (2008, pp. 174–182) analysis of another Radio 5 Live 'discussion' of the views of a racist academic make disturbing reading, but, from my perspective are indicative of a deeply racist society rather than suggestive of 'white supremacy'. I agree totally with Gillborn (2008, p. 181) that '*there is simply no such thing as entirely free speech*', and that 'the supposed neutrality and liberalism of free speech doctrines actually work in the favour of the already powerful' (ibid.). Gillborn (ibid.) notes the prohibitions on libel, defamation, copyright, and incitement to 'terrorism'. It is worth noting here the oft-quoted judgment by Oliver Wendell Holmes, Jr. in the U.S. Supreme Court case *Schenck v. United States,* in 1919 when Holmes pronounced on the limits of free speech under the terms of the First Amendment of the U.S. Constitution. Holmes wrote:

> The most stringent protection of free speech would not protect a man falsely shouting fire in a theater and causing a panic. [...] The question in every case is whether the words used are used in such circumstances and are of such a nature as to create a clear and present danger that they will bring about the substantive evils that Congress has a right to prevent. (Wikipedia, 2008)

Moreover, I have always argued that the right not to be on the receiving end of verbal violence such as racism (a violence which sometimes has the effect of enticing others to physical violence) is more important than the right of the racist to incite such hatred.

The conclusion in Gillborn's chapter that the 'rules of the game are defined by, and for, White people [in the] interests of White people' (Gillborn, 2008, p. 182) should by now be as predictable as my response of 'not White people' but 'the (predominantly white) racist capitalist state'.

On Conspiracy

In the Conclusion to the book, Gillborn returns to the Chronicle featuring The Professor and Steve. The topic of their conversation is 'conspiracy' with Steve, the law student, referring to 'concert of action for a common purpose' (Gillborn, 2008, p. 192), and arguing that the chapters of the Professor's book [Gillborn, 2008] describe 'concerted practice...always to the benefit of the racist status quo' (Gillborn, 2008, p. 192)—'a web of actions by teachers, policymakers, right-wing commentators, uncritical academics and the media—all working in one direction, day after day and to incredibly powerful effect' (Gillborn, 2008, p. 192). This seems to me to be a pretty accurate description of how the system works, although, for me, the beneficiary of this 'common purpose' is not 'Whiteness', but the capitalist state. For my part, how the system works is best explained, as I have argued in this volume, as the successful interpellation of subjects (see also Cole, 2008g, 2008h). Steve further suggests that there is a hub-and-spoke conspiracy which links people together 'by the shared "common-sense" assumptions and actions that characterize them' (Gillborn, 2008, p. 193). I would agree with that people are linked by such assumptions and actions, and again, I would attribute this to interpellation. Moreover, to repeat, whereas for Steve and the Professor (and Gillborn), the central hub is 'Whiteness' (ibid.), for me it is the racist (predominantly white) capitalist state. All white people do not have, as I argued in chapter 2 of this volume, the 'cultural and economic dominance' (ibid.) afforded to them by Steve, The Professor and Gillborn.

And Finally...On 'Struggling Where We Are' against 'the Powers that Be'

Gillborn's (2008) book ends with an Appendix, in which Gillborn's final message is that we must 'struggle where we are' (Gillborn, 2008, p. 202). What Gillborn has in mind is that 'regardless of our institutional location (in a school, university, school district, factory, community group or political party)', and regardless of our identity, we need to struggle against racism (p. 202). While Marxists would prioritize struggle at the point of production as being most effective, because it is here that capitalism sustains and reproduces itself (whether that is the production of cars or the production of knowledge), there would be no disagreement with the notion that we struggle where we are.

As to whom we struggle against, citing Bell, 1992b, pp. 198–199, Gillborn states that we need to 'remind the powers that be that out there are persons like us who are not only not on their side but determined to stand in the way' (Gillborn, 2008, p. 203). For Critical Race Theorists 'the powers that be' are encapsulated by the concepts of 'White powerholders' and 'white supremacy'; for Marxists these powers, as I have repeatedly argued, are centered in the world system of racialized (and gendered) neoliberal global capitalism and imperialism. In chapter 6, I will turn to an examination of this world system itself. I will then in chapter 7 assess the contributions of Marxism to a twenty-first century socialism, before concluding the book (chapter 8) with some suggestions for classroom practice from both CRT and Marxism.

Chapter 6

Neoliberal Global Capitalism and Imperialism in the Twenty-first Century

In this chapter, I examine the issues of *capitalism, globalization, neoliberalism* and *globalization and global environmental destruction* both conceptually and materially. I then relate these issues to *globalization and the U.S. Empire*, assessing some comments on globalization from two leading Critical Race Theorists, and some perspectives from transmodernism and postmodernism on globalization and the U.S. Empire, before outlining my preferred Marxist analysis.[1] I conclude the chapter with some comments on the imperial occupation of Iraq five years on.

Capitalism

Capitalism is, from a Marxist perspective *by definition*, a system in which a minority (the capitalist class) exploits the majority (the working class) by extracting surplus value from their labor power (this is developed in the next chapter). It is a system without morality and without shame. It is a system of intense and relentless exploitation. As Michael Parenti (1998, pp. 84–85) has put it:

> Capitalism is a system without a soul, without humanity. It tries to reduce every human activity to market profitability. It has no loyalty to democracy, family values, culture, Judeo-Christian ethics, ordinary folks, or any of the other shibboleths mouthed by its public relations representatives on special occasions. It has no loyalty to any nation; its only loyalty is to its own system of capital accumulation. It is not dedicated to 'serving the community'; it serves only itself, extracting all it can from the many so that it might give all it can to the few

Capitalism has an inbuilt tendency to constantly expand. Marx and Engels recognized its preeminent global character over one hundred and fifty years ago. As they put it in *The Communist Manifesto*, when describing capitalism's development:

> The markets kept ever growing, the demand ever rising...The place of manufacture was taken by the giant, Modern Industry, the place of the industrial

middle class, by industrial millionaires...Modern industry has established the world-market. The need of a constantly expanding market for its products chases the bourgeoisie over the whole surface of the globe. It must nestle everywhere, settle everywhere, establish connexions everywhere...In one word, it creates a world after its own image. (Marx and Engels (1847) [1977a], pp. 37–39)

Glenn Rikowski (2001, p. 21) has clarified what is entailed in this expansion, a process which takes three main forms: first, spatial (globalization), as capital occupies all known sociophysical space (including outside the planet)—this is *extension*; second, capital expands as the differentiated form of the commodity, creating new commodities—this is *differentiation*; third, it expands through *intensification* of its own production processes (Rikowski, 2001, p. 21). Capitalism is thus a thoroughly dynamic system.

In its inherent need to extract more and more surplus value, capital is also out of control. As Rikowski has argued:

Capital moves, but not of its own accord: the mental and physical capabilities of workers (labour-power) enable these movements through their expression in labour. The social universe of capital then is a universe of constant movement; it incorporates and generates a restlessness unparalleled in human history...It is set on a trajectory, the 'trajectory of production'...powered not simply by value but by the 'constant *expansion* of surplus value' (Postone, 1996, p. 308 [Rikowski's emphasis])...It is a movement out of control. (Rikowski, 2001, p. 11)

Chris Harman (2008, p. 11) has described twenty-first century global capitalism, which rests on the unplanned interaction of thousands of multinationals and twenty or so nation states, as resembling 'a traffic system without lane markings, road signs, traffic lights, speed restrictions or even a clear code that everyone has to drive on the same side of the road'.

Globalization

Globalization always has been and is a central feature in the maintenance and parasitic growth of capitalism. However, it became one of the orthodoxies of the 1990s and continues to hold sway into the twenty-first century, as a *new* phenomenon. Its premises are that in the face of global competition, capitalist organizations are increasingly constrained to compete on the world market. Its argument is that, in this new epoch, these organizations can only do this in so far as they become multinational corporations and operate on a world scale, outside the confines of nation states (Harman, 1996). The argument continues: this diminishes the role of the nation-state, the implication being that there is little, if anything, that can be done about it. Capitalists and their allies, particularly pro-capitalist politicians, insist that, since globalization is a fact of life, it is incumbent on workers, given this globalized market, to be flexible in their approach to what they do and for how long they do it; to accept lower wages; and to concur with the restructuring and diminution of welfare states (Cole, 2008d). It is important to stress the *ideological* nature of this scenario, and to note that, while globalization is taking new forms, essentially it is as old as capitalism itself. Marxists are not only interested in processes of modern globalization, but are also interested in this ideological element which furthers the interests of capitalists and their political supporters (for an analysis, see Hill, 2003; Hill, 2009a, 2009b; Hill and Kumar, 2009; Hill and Rosskam, 2009; see also Cole, 1998a, 2005), of the way in which it is used to mystify the populace as a whole and to stifle action by the Left in particular (e.g., Murphy, 1995; Gibson-Graham, 1996; Harman, 1996; Meiksins Wood, 1998).

Critical Race Theorists, Delgado and Stefancic (2001, p. 111) argue that globalization 'is very much in the forefront of critical race theory'. They then proceed with a fundamentally Marxist analysis of the direction globalization is taking in the twenty-first century: a globalizing economy removes manufacturing jobs from the inner city; it creates jobs in the knowledge economy, for which minorities have little training; the sweatshops and other exploitative conditions it creates afflict poor people of color, many of them women in developing countries, which were formally colonized; globalization concentrates capital in the hands of an elite class who refuse to share it. The Marxist-inspired explanation continues: globalization creates alliances of United States and 'third world' workers against American corporations; it facilitates mobilization of labor unions; and protests against the WTO ensue. The reason wages are low and the new jobs are attractive, they continue (pp. 111–120) is because United States and European colonialism has robbed the former colonies of their natural wealth, stifled the development of local leaders and conspired with right-wing dictators to keep the people poor and unorganized. 'If the materialist wing (see chapter 1 pp. 21–22 of this volume for a discussion) of critical race theory is right', they state, 'domestic minorities have suffered at the hands of very similar forces' [I would add white workers have suffered too]. In classic Marxist fashion they conclude, '[domestic minorities]' fates are linked with those of their overseas counterparts, since capitalists can always use the threat that investments will relocate overseas to defeat unions, workplace regulations, welfare, and other programs of interest to U.S. minorities' [again I would want to add, 'and white workers'].

Neoliberalism

Globalization in the twenty-first century is pre-eminently neoliberal, the adoption of which has given a major boost to globalization, both de facto and ideologically. Martinez and García (2000) have identified five defining features of the global phenomenon of neoliberalism (see also Hill, 2003, 2004, 2005a, 2005b):

1. *The Rule Of The Market*
   - the liberation of 'free' or private enterprise from any bonds imposed by the state no matter how much social damage this causes;
   - greater openness to international trade and investment;

- the reduction of wages by de-unionizing workers and eliminating workers' rights;
- an end to price controls;
- total freedom of movement for capital, goods and services.

2. *Cutting Public Expenditure*
   - less spending on social services such as education and health care;
   - reducing the safety-net for the poor;
   - reducing expenditure on maintenance, for example, of roads, bridges and water supply.

3. *Deregulation: reducing government regulation of everything that could diminish profits*
   - less protection of the environment;
   - lesser concerns with job safety.

4. *Privatization: selling state-owned enterprises, goods and services to private investors, for example,*
   - banks;
   - key industries;
   - railroads;
   - toll highways;
   - electricity;
   - schools;
   - hospitals;
   - fresh water.

5. *Eliminating the Concept of 'The Public Good' or 'Community'*
   - replacing it with 'individual responsibility';
   - pressuring the poorest people in a society to by themselves find solutions to their lack of health care, education and social security.

Global neoliberalism was given a major boost in 1994, with the signing of the General Agreement on Trade in Services (GATS) at the World Trade Organization (WTO). The aim of this agreement, which came into force in January 1995, is to remove any restrictions and internal government regulations in the area of service delivery that are considered to be 'barriers to trade'. The list of services of the GATS includes 12 types, subdivided in many others:

1. Business (accounting, computer science and related subjects, legal, marketing and correlated, medical and dental services, architecture, etc);
2. Communication (telecommunication, mail, audiovisual, radio, motion picture etc);
3. Construction and related engineering services;
4. Distribution (franchising, retail and wholesale, etc);
5. Education (primary, secondary, higher, adult education and others);
6. Environmental (sewage, sanitation, disposal, etc);
7. Financial (insurances, banking, leasing, asset management, etc);
8. Health and related social services (hospital, other human health services, social, etc);

9. Tourism and travel related (hotel, restaurant, travel agencies, etc);

10. Recreational, Cultural and Sporting (news agency, libraries, archives, museums, theatre, sports, etc.;

11. Transports (maritime, aerial, railway, railroad, passenger, freight, maintenance and repair, towing, pipelines, warehouses, etc);

12. 'Other services not mentioned in any other place.' (WTO, 2003, cited in de Siqueira, 2005)

Since February 2000, negotiations have been underway in the WTO to expand and 'fine-tune' the GATS. As GATSWatch (undated) has pointed out, these negotiations have aroused concern worldwide. A growing number of local governments, trade unions, non-governmental organizations (NGOs), parliaments and developing country governments are criticizing the GATS negotiations and call for a halt on the negotiations. Their main points of critique are as follows:

- Negative impacts on universal access to basic services such as healthcare, education, water and transport.
- Fundamental conflict between freeing up trade in services and the right of governments and communities to regulate companies in areas such as tourism, retail, telecommunications and broadcasting.
- Absence of a comprehensive assessment of the impacts of GATS-style liberalization before further negotiations continue.
- A one-sided deal. GATS is primarily about expanding opportunities for large multinational companies (GATSWatch undated; see also Devidal, 2009; Verger and Bonal, 2008; Waghorne, 2008).

Globalization and Global Environmental Destruction

The unrelenting abuse of nature, viewed as a resource to plunder by global neoliberal capitalism, has had disastrous consequences. Millions of poor people have been driven off their land, while whole areas of agricultural land have been damaged, and rain forests destroyed by mining, logging and oil companies. Our health is seriously at risk by the food we eat, genes are being engineered and modified, and 'global warming' is threatening the survival of life on the planet. Elsewhere (Cole, 2008d, pp. 90–96; see also Feldman and Lotz, 2004) I have dealt with the effects of environmental destruction under the following headings: *Unhealthy Food; Genetic Modification;* the *Destruction of Resources;* and *Climate Change.* I have argued that the food that we eat in 'developed' countries is unhealthier than ever before, and that it is estimated that 70 percent of the £20 million global annual food advertising budget is used to promote (unhealthy) soft drinks, sweets and snacks (Feldman and Lotz, 2004, p. 129).

I further noted that the last twenty-five years or so has seen a dramatic extension and deepening of global capitalism's penetration of nature for profit. For example, genetic modification, having first occurred in 1973, is an unprecedented incursion. Moreover, this knowledge is being privatized through patents on genes (Feldman and Lotz, 2004, p. 118). Paul Gilroy (2004, p. 84) has described these developments as the 'corporate control of the substance of life itself', 'linking 'the colonization of territory and human beings with the colonization of all life'.

Jeremy Rifkin, 1999, cited in Feldman and Lotz, 2004, p. 137, has summed up the dangers of genetic engineering as a whole, where:

> [a] handful of corporations, research institutions and governments could hold patents on virtually all 100,000 genes that make up the blueprint of the human race, as well as the cells, organs, and tissues that comprise the human body. They may also own similar patents on thousands of micro-organisms, plants and animals, allowing them unprecedented power to dictate the terms by which we and future generations will live our lives

With respect to the destruction of resource, I pointed out how intensive farming in the last 60 years and the turn to industrialized agriculture under current globalization have resulted in ecological catastrophe. Of particular concern is the destruction of rainforests, home to more species of plants and animals than the rest of the world put together. The drilling and production of oil is also a great threat to large areas of rainforests. Burning oil and other fossil fuels pollutes the atmosphere, and contributes to global warming and climate change, one of the greatest threats to the survival of the all the inhabitants, and indeed all living things on our planet.

Glaciers in Greenland are slipping into the sea at a rate that doubled between 1996 and 2000, and the Antarctic ice cap, which holds 70 percent of the world's water, is now losing water at the same rate as Greenland (Ward, 2006, p. 12). The causal role of neoliberal global capitalism in global warming is indisputable. An annual growth rate (GNP) of 3 percent (the accepted rate for the developed world) means that production is doubled every 24 years, and there is a close correlation between GNP and the rate of increased fossil fuel use (Kinnear and Barlow, 2005). As Phil Ward (2005, p. 14) puts it, 'the capitalist system...is incapable of downsizing except by means of destructive slump or war'. As argued earlier, capitalism is out of control 'set on a trajectory, the "trajectory of production"...powered not simply by value but by the "constant *expansion* of surplus value"' (Postone, 1996, p. 308, cited in Rikowski, 2001, p. 11). (Rikowski's emphasis)

Petroleum is the main fuel used by consumers. The connection between increased fossil fuel use and imperialist adventures in oil-rich countries is an obvious one. One of the primary reasons for U.S. imperial expansion is, of course, to control access to, and the marketing of oil (the other being U.S. capitalist hegemony). This, in turn, creates further environmental degradation and destruction, both in the United States, and worldwide. I will now consider the role of the 'New Imperialism' in the twenty-first century, and, in the last chapter of this volume, will argue the case for a study of imperialisms to be a central feature of the curriculum.

Globalization and the U.S. Empire

Ellen Meiksins Wood (2003, p. 134) has captured succinctly globalization's current imperialist manifestations:

> Actually existing globalization...means the opening of subordinate economies and their vulnerability to imperial capital, while the imperial economy remains sheltered as much as possible from the adverse effects. Globalization has nothing to do with free trade. On the contrary, it is about the careful control of trading conditions in the interest of imperial capital (cited in McLaren and Farahmandpur, 2005, p. 30)

While globalization is used to further the interests of capitalists and their supporters per se, it is often similarly used *ideologically* to justify the New Imperial Project. On September 17, 2002, a document entitled *National Security Strategy of the United States of America* (*NSSUSA*) was released which laid bare U.S. global strategy in the most startling terms (Smith, 2003, p. 491). As transmodernist, David Geoffrey Smith points out, the Report heralds a 'single sustainable model for national success: freedom, democracy and free enterprise'. Europe is to be kept subordinate to, and dependent on, U.S. power, NATO is to be reshaped as a global interventionist force under U.S. leadership, and American national security is claimed to be dependent on the absence of any other great power. The Report also refers to 'information warfare', whereby deliberate lies are spread as a weapon of war. Apparently, a secret army has been established to provoke terrorist attacks, which would then justify 'counter attack' by U.S. forces on countries that could be announced as 'harboring terrorists' (The Research Unit for Political Economy (RUPE), 2003, pp. 67–78, cited in Smith, 2003, pp. 491–492).

While the *NSSUSA* states that American diplomats are to be retrained as 'viceroys' capable of governing client states ((RUPE), 2003 cited in Smith, 2003, p. 491), the New Imperialism, in reality, no longer seeks direct territorial control of the rest of the world, as did British Imperialism for example, but instead relies on 'vassal regimes' (Bello, 2001, cited in Smith, 2003, p. 494) to do its bidding. This is because capital is now accumulated via the control of markets, rather than by *sovereignty* over territories.

Writing from a liberal perspective, Michael Lind (2004, p. 5) points out that this does not stop many neo-Conservatives in the United States hankering after British Imperialism (and in particular the young Winston Churchill) as their model. British neo-Conservative popular historian and TV presenter Niall Ferguson, for whom the British Empire was relatively benevolent, has similar views. In a speech in 2004, he argued that the American Empire which 'has the potential to do great good' needs to learn from the lessons of the British Empire. First it needs to export capital and to invest in its colonies; second, people from the United States need to settle permanently in its colonies; third, there must be a *commitment* to imperialism; fourth there must be collaboration with local elites. Success can only come, he concludes if the Americans are prepared to *stay* (Ferguson, 2004). George Bush and Tony Blair were, of course, pivotal in extending and consolidating U.S. imperialism. Ferguson (2005) argued that Bush is an 'idealist realist' who is 'clearly open to serious intellectual ideas'. Bush is a realist because he believes that power is 'far more important than law in the relations between states', and an idealist because he wants to spread 'economic and political freedom around the world'. Bush, he goes on, has picked up two main ideas from the academy, namely that free markets accelerate economic growth which makes democracy more likely to succeed, and democracies are 'much less likely to make war than authoritarian regimes'. Ferguson then offers the President a further idea. It helps to think of the U.S. Empire (Ferguson's words not mine) 'as a kind of sequel to the British Empire'. The lesson to be learnt from that Empire is the need to stay longer. 'Elections are not everything' and the danger posed to liberty in the United States, and on the imperial front, he concludes, is less worrying than 'a decline in US power...surely something about which idealists and realists can agree'.

Wall Street Journalist, Max Boot has gone so far as to state that 'Afghanistan and other troubled lands today cry out for the sort of enlightened foreign administration provided by self-confident Englishmen in jodhpurs and pith helmets' (cited in Smith, 2003, p. 490) (see pp. 57–60 and p. 110 of this volume for a discussion of the imperialist views of Barack Obama; see also the Postscript to this volume).

Enfraudening and Enantiomorphism: A Transmodern Perspective

Information warfare is a key imperialist strategy and *modus operandi* of capitalism; so is 'enfraudening the public sphere'. 'Enfraudening the public sphere' is a term coined by David Geoffrey Smith (Smith, 2003, p. 488–489) to describe 'not just simple or single acts of deception, cheating or misrepresentation' (which may be described as 'defrauding'), but rather 'a more generalized active conditioning of the public sphere through systemized lying, deception and misrepresentation'.

The major strength of transmodernism, I would argue, lies in its argument that European philosophers still are not facing the historical responsibilities of their legacies (Smith, 2004, p. 644). As I argued elsewhere (Cole, 2008d), transmodernism makes an important contribution to an understanding of the legacy of the European invasion of the Americas because it reveals how the imperialism in which contemporary U.S. foreign policy is currently engaged has a specific and long-standing genealogy.

Smith (2003, p. 489) argues that the Bush Administration's 'war on terror' was used to veil long-standing, but now highly intensified, global imperial aims. Following McMurtry (1998, p. 192), he suggests that, under these practices, knowledge becomes 'an absurd expression' (Smith, 2003, p. 489). Again, following McMurtry (2002, p. 55), Smith (2003, pp. 493–494) argues that the corporate structure of the global economy (dominated by the United States, particularly through its petroleum corporations) 'has no life co-ordinates in its regulating paradigm' and is structured to misrepresent its indifference to human life as "life-serving"'. Thus we have terror in the name of anti-terrorism; war in the name of peace seeking. Accordingly, U.S. secretary of state, Colin Powell (2003) was able to declare with a straight face and in a matter-of-fact tone that the 'Millennium Challenge Account' of the Bush administration was to install 'freely elected democracies' all over the world, under 'one standard for the world' which is 'the free market system...practiced correctly' (cited in Smith, 2003, p. 494). This provides the justification for the slaughter of hundreds of thousands of Iraqi children since 1990 through NATO bombing and the destruction of the public infrastructure (water, healthcare, etc.). This slaughter has, of course, taken on a new dimension since the March 2003 invasion and occupation of Iraq. Such justification is also given for the destabilization of democratically elected governments throughout Latin America, Africa and Asia (Smith, 2003, p. 494).

Smith (2003, p. 494) describes this rhetorical process as enantiomorphic—whereby a claim is made to act in a certain way, when one actually acts in the opposite way. Enantiomorphism reached its zenith, I would argue, in the absurd claim nurtured by Bush and Blair that the invasion and occupation of Iraq was necessary because Saddam Hussein had weapons of mass destruction, which he was going to use on the West. There were also *reasonable* claims made that he tortured his people and was anti-democratic. The Americans and their allies were going there, we were told, to find the weapons of mass destruction, stop the people being tortured, and bring democracy. The reality is, of course, that not only did Saddam have no weapons of mass destruction (it is the Americans who have such weapons, and remain the only country that has dropped atomic bombs in warfare) but the Americans and the British have continued the torture (see later in this chapter); and upheld the lack of democracy.[2]

A Postmodern Fantasy

In Cole, 2008d, pp. 98–100, I also discussed the 'postmodern fantasy' of Robert Cooper (2002, p. 5).[3] Briefly, Cooper argues that postmodern imperialism takes two forms. The first is the voluntary imperialism of the global economy, where institutions like the IMF and the World Bank provide help to states 'wishing to find their way back into the global economy and into the virtuous circle of investment and prosperity' (ibid.). If states wish to benefit, he goes on 'they must open themselves up to the *interference* of international organizations and foreign states' (ibid.) (my emphasis). Cooper (ibid.) refers to this as a new kind of imperialism, one which is needed and is acceptable to what he refers to as 'a world of human rights and cosmopolitan values': an imperialism 'which, like all imperialism, aims to bring order and organisation' [he does not mention exploitation and oppression] 'but which rests today on the voluntary principle'. While '[w]ithin the postmodern world, there are no security threats'...'that is to say, its members do not consider invading each other' (p. 3), that world, according to Cooper has a right to invade others. The 'postmodern world' has a right to pre-emptive attack, deception and whatever else is necessary.

The second form of postmodern imperialism Cooper calls 'the imperialism of neighbours' (Cooper has in mind the European Union), where instability 'in your neighbourhood poses threats which no state can ignore'. It is not merely soldiers that come from the international community; he argues, 'it is police, judges, prison officers, *central bankers* and others' (my emphasis). Between 1999 and 2001, Cooper was Tony Blair's head of the Defence and Overseas Secretariat, in the British Cabinet Office.

Transmodern 'Narcissism' or Racializing the Other: A Marxist Analysis

While Smith's arguments on enfraudening and enantiomorphism are convincing, and in centering on the role of ideology, in essence Marxist, I have problems with the vague transmodern notions of 'narcissism'[4] in explaining the source of Western violence directed against the Other. As Paul Warmington (2006, personal correspondence) has pointed out, the transmodern notion of 'narcissism' is problematic for Marxists. First, it represents essentialist notions of 'kinship'; a natural tendency to align oneself with one's 'own kind'. Second, because its psychosocial gloss does not take account of Marxist understandings of the material base of discourse, it inverts the historical relationship between imperialism and Otherness. Far from deriving from a narcissistic *alignment* with one's own kind and antipathy to the Other, I would argue, following Warmington (ibid.), that the western violence that enforced capitalist imperialism (from the sixteenth century onward) entailed a conscious and *strategic* (and traumatic) *alienation* from other nations (as well as from the west's own emergent liberal-democratic values). This historically specific alienation was achieved through contrived 'racial', cultural and spatial distinctions that served to mask the key contradictions of imperialist production. 'Race' and racialization were key factors here.

As I agued in chapter 5 of this volume, the rhetoric of the purveyors of dominant discourses aims to shape 'common sense discourse' into formats which serve their interests. Underlining the fragmentary and incoherent role of 'common sense' in connecting racialization to popular consciousness, Peter Fryer (1988), outlines the following argument. Modern racist ideology emerged with and from the Atlantic slave trade (which predated the 'mature' colonialism of the Indian sub-continent by 150 years) and was anomalous in that:

- at the point when western European production was shifting towards free labor and was shifted by technological advances, it made itself increasingly reliant on a backward form of production, that is, chattel slavery
- at the point at which the emergent Enlightenment began to posit notions of individual freedom, the west embarked on conquest and enslavement, in order to secure servile labor systems (first in America, later in colonial Asia and Africa).

As Warmington (2006, personal correspondence) argues, racialization can be seen, therefore, as a project to rhetorically 'resolve' these contradictions, not merely to justify them in the sense of papering over their cracks but to construct a racialized 'justice' upon which to build brutal, servile production systems. In short, if liberty and the Enlightenment were morally and ideologically correct then they must necessarily be extended to all humanity (and this was the view of some dissident voices in the west). However, this extension was clearly impossible (both at home and abroad—but especially abroad in those continents that were, as Du Bois pointed out, being subjected to conquests that made them bear the largest burden of global pauperization, conquests involving unprecedented levels of violence and displacement). Thus an ideology was required that placed the slave labor force outside the bounds of humanity and therefore outside the 'human rights' being tentatively proclaimed in Europe (clearly this ideology also infused racial folklore). Fryer (1988, p. 63) quotes Genovese and Genovese: '[The rising capitalist] class *required* a violent racism not merely as an ideological rationale but as a psychological imperative.'

In Cole, 2006a I addressed the origins of the New Imperialism, how 'the eclipse of the non-European' following the European invasion of 1492, consolidated by subsequent invasions and conquests, unleashed racialized capitalism, often gendered, on a grand scale. The expansion of capital entailed not only the beginnings of the transatlantic slave trade, but also the attempted enslavement, the massacre, and the seizing of the land of indigenous peoples, both local and adjacent. Its legacy today includes a very high and disproportionate suicide rate for Native Americans in general, and continuing attacks on the reproductive rights of Native American women; the 'prison industrial complex'—a legacy of slavery—where 'people of color' are disproportionately represented; human rights abuse at U.S. borders; and continuing segregation in U.S. cities. Racialized notions of 'like' and 'Other' ('black' and 'white', 'civilized', and 'savage') are *ends* (or mediators), the starting point being shifts in production (slavery and colonialism's forms of sixteenth- to early-twentieth-century globalization). 'Otherness' was a strategic, violent creation.

Once groups have become racialized via 'common sense', for example, as 'savages' in the case of indigenous peoples, or sub-human and genetically inferior, as in the case of African slaves, genocide becomes less problematic (Cole, 2006a; see also McLaren, 1997). In a similar fashion, once Muslims are racialized as the Other (and the 'war on terror' knows no bounds) torture, humiliation and other human rights abuses, to which Guantanamo Bay and Abu Ghraib bear witness, becomes routine practice. Such practice is not confined to these locations. Former detainee, Moazzam Begg (Begg and Brittain, 2006) for one, recalls abuse in U.S. and British military prisons in Pakistan, Afghanistan, Iraq and Egypt, as well as in Guantanamo Bay (see also Campbell and Goldenberg, 2004).

Such treatment is sustained by racialization. Indeed, the historic *a priori* racialization of Native Americans and African Americans as sub-human, and

Muslims as sub-human and terrorists serves to legitimate and facilitate their massacre, enslavement, torture, rape, humiliation and degradation. In the current era, global imperialist abuse involves psychological as well as physical abuse, with detainees denied halal meat, for example. In addition, sexual torture has been revealed as having occurred on a massive scale, and as having apparently been developed by intelligence services over many years. In particular the humiliation of the body stands in stark contrast to the Muslim importance of covering, and not exposing flesh. Such abuse has also involved sexual humiliation. In 2003, U.S. soldier, Lynndie England serving at the Abu Ghraib camp in Iraq was charged with abusing detainees and prisoners by forcing them to lay in a naked pyramid with an aim to humiliate. Photographs taken by U.S. military also showed Lynndie England holding a leash attached to the neck of a naked man on the floor (Sands, 2008, p. 23), while another showed a prisoner with wires attached to his fingers, standing on a box with his head covered (ibid.) *BBC News* (2004) reported that there 'were numerous incidents of sadistic and wanton abuse. ...Much of the abuse was sexual, with prisoners often kept naked and forced to perform simulated and real sex acts'.

Torture techniques, approved by Donald Rumsfeld, and endorsed by George Bush were in three categories: Category 1 comprised yelling and deception; Category II included 'humiliation and sensory deprivation, including stress positions, such as standing for a maximum of four hours; isolation; deprivation of light and sound; hooding; removal of religious and all other comfort items; removal of clothing; forced grooming, such as shaving of facial hair; and the use of individual phobias, such as fear of dogs, to induce stress' (Sands, 2008, p. 21). Category III techniques were to be used for a very small percentage of detainees, 'the most uncooperative (said to be fewer than 3%) and exceptionally resistant individuals—and required approval by the commanding general at Guantánamo' (ibid.). There were four techniques in the last category: 'the use of "mild, non-injurious physical contact", such as grabbing, poking and light pushing; the use of scenarios designed to convince the detainee that death or severely painful consequences were imminent for him or his family; exposure to cold weather or water; and, finally, the use of a wet towel and dripping water to induce the misperception of suffocation' (ibid.).

According to Sands (ibid.) the pattern at Guantanamo was always the same, and consisted of:

> 20-hour interrogation sessions, followed by four hours of sleep. Sleep deprivation appears as a central theme, along with stress positions and constant humiliation, including sexual humiliation. These techniques were supplemented by the use of water, regular bouts of dehydration, the use of IV tubes, loud noise (the music of Christina Aguilera was blasted out in the first days of the new regime), nudity, female contact, pin-ups. An interrogator even tied a leash to [one detainee], led him around the room and forced him to perform a series of dog tricks. He was forced to wear a woman's bra and a thong was placed on his head.

What seemed to unite torture at Guantanamo Bay and Abu Ghraib was 'humiliation, stress, hooding, nudity, female interrogators, shackles, dogs' (Sands, 2008, p. 23). Such sexualized abuse is part and parcel of the racialization of the Other in the pursuit of hegemony and oil (see the appendix to this chapter). Global rule and the New Imperialism are, of course, first and foremost, about global profits. This connection to capital, national and international is outside the remits of both transmodernism and CRT, thereby rendering their use as a tool for analysis significantly lacking.

Racialization, under conditions of imperialism is fired by what Dallmayr (2004, p. 11), has described above as 'the intoxicating effects of global rule' that anticipates 'corresponding levels of total depravity and corruption among the rulers'.

The racialization of the Other provides a more convincing explanation of the justification of conquest and enslavement by the West and of 'The New Imperialism' than the transmodern exaltation of basic narcissism as a causal factor. The concept of 'narcissism' is unconvincing because it *starts* from the opposition of 'like' and 'Other', and because it conflates ahistorical notions of 'Otherness' with historically specific forms of racialization. While, as noted in chapter 1, Critical Race Theorists like Ladson-Billings and Tate (1995, p. 55) note the plunder of native Americans land through military conquest of the Mexicans to the construction of Africans as property, they are unable within their own frames of reference, *without resorting to Marxist analysis*, to relate all this to capitalism. As Darder and Torres (2004, p. 99) put it, the efforts of Critical Race Theorists:

> to explore the ways in which socioeconomic interests are expressed in the law or education are generally vague and undertheorized. Because of this lack of a theoretically informed account of racism and capitalist social relations, critical race theory has done little to further our understanding of the political economy of racism and racialization.

For Marxists, the historical and contemporaneous racialization of the Other via 'common sense' must be connected historically and contemporaneously to changes and developments in the mode of production. Indeed, as I have tried to demonstrate in this book, for Marxists, an analysis of racism *begins* with the capitalist mode of production, with social class and with class struggle (Darder and Torres, 2004, p. 99). In the current era, capital is preeminently under U.S. control. We live in a world, much of which is increasingly at the beck and call of the White House, and of the diktats of the New Imperialism, where globalization is portrayed as inevitable, and imperialistic designs are masked as 'the war against terror' and the promotion of democracy.

The Occupation of Iraq Five Years On

At the time of writing (March, 2008) U.S. military deaths in Iraq has reached 4,000 (Kay, 2008). At the same time, at least 60,000 more troops have been wounded, and many thousands more American soldiers and Marines have come back with severe psychological problems (World Socialist Web Site (WSWS) Editorial Board, 2008a).

However, even this tragedy pails in significance compared to the estimated 1 million plus Iraqis who have been killed, while a further 4 million have become refugees (ibid.). The war has been described by the WSWS Editorial Board (ibid.) as 'the greatest geo-political disaster in American history'. As the point out (ibid.):

> The war's costs, in terms of both US imperialism's global position and sheer dollar amounts, have eclipsed the immense damage wrought by the protracted intervention in Vietnam nearly four decades ago. It has already lasted longer than the American Civil War, World War I, World War II and the Korean War. Even in Vietnam, after five years of major troop deployments, the withdrawal of American forces had already begun.

They quote The International Tribunal at Nuremburg that convicted the leaders of the Third Reich:

> War is essentially an evil thing. Its consequences are not confined to the belligerent states alone, but affect the whole world. To initiate a war of aggression, therefore, is not only an international crime, it is the supreme international crime, differing from other war crimes in that it contains within itself the accumulated evil of the whole. (ibid.)

Incredibly a poll conducted in March 2008 for the BBC, ABC News in the U.S., German and Japanese television found that nearly half of the residents of Baghdad said that at least one family member had been killed since the occupation began. The poll also revealed that over 70 percent of Iraqis want U.S. troops out of their country (ibid.).

Essential infrastructure remains devastated, since the American high explosives of five years ago, as well as the previous years of punishing sanctions. This means that the population is deprived of electricity, fuel, clean water, sanitary facilities and garbage collection. Moreover, the killing of over 600 doctors and medical professionals and the flight of thousands of others, together with severe shortages in medicine and equipment, have left Iraq's health sector in a state of collapse (ibid.).

As far as the costs to American society are concerned, it is estimated that the occupation consumes some $12 billion a month. A report by the Joint Economic Committee of Congress estimated that the war thus far has cost an average American family of four $16,900, an amount projected to rise to $37,000 by 2017. This huge amount of money has been diverted from pressing social needs in the United States, with the massive expenditures contributing significantly to a raging financial crisis that threatens to plunge the economy into a depression (ibid.).

Despite all this, Vice President Dick Cheney, during an unannounced visit to Baghdad, called the five-year war a 'successful endeavor' that 'has been well worth the effort', while in a video conference in March 2008 with U.S.

military personnel in Afghanistan, President Bush declared himself envious of those fighting in America's colonial-style wars, calling it 'a fantastic experience' and 'in some ways romantic' (ibid.). The reality is that five years after a U.S. invasion that was expected by its organizers to swiftly replace the government of Saddam Hussein with a stable U.S. client regime, 160,000 U.S. troops remain deployed in the country and no area can be claimed to be fully secure (ibid.).

The enantiomorphic claims about weapons of mass destruction are now universally recognized to be false, as are claimed links between Saddam Hussein and Al Qaeda, both of which proved to be non-existent.

As the WSWS Editorial Board, 2008a point out:

> The Bush administration, with the complicity of congressional Democrats, sought to exploit the fears and political confusion in the wake of the September 11 terrorist attacks to implement long-prepared plans to seize control of a country holding the word's second-largest proven oil reserves and turn it into a platform for the extension of US military power throughout the region.

Despite all this, there was and continues to be mass popular opposition to the war. The American people by a large margin have come to oppose the war, yet it continues unabated, and the president who launched, who is despised by millions and retains the support of less than a third of the population, retains undiminished power to pursue a policy of unrestrained militarism (ibid.) As the WSWS Editorial Board (ibid.) note, '[n]othing could expose more thoroughly the undemocratic character and political rot that pervade the entire governmental system within the United States'. They sum up the current realities of twenty-first century U.S. imperialism:

> The global eruption of American militarism and the crisis of US and world capitalism are inextricably linked. In the final analysis, the wars in Iraq and Afghanistan, and the threat of a new war against Iran, are a product of the attempt by the US ruling class to maintain the hegemonic position of US capitalism by military force, under conditions in which it can no longer do so by virtue of its economic weight. The most important war aims of Washington are to establish a stranglehold over the oil resources of the Middle East and Central Asia, in order to gain a decisive strategic advantage over its economic rivals in Europe and Asia. (ibid.)

However, it is an imperialism in decline. Hence despite the failures of Iraq, desperation over its threatened loss of hegemony is pushing Washington towards new confrontations with enemies ranging from China to Russia to Venezuela.

It is, of course, the working class who suffer, as the 'financial elite's policy of using military force to gain control of world markets is pursued at the direct expense of the masses of working people, who are paying for it through attacks on their jobs, living standards and basic democratic rights' (ibid.).

As WSWS writer David North (2003) predicted accurately at the start of the war:

> Whatever the outcome of the initial stages of the conflict that has begun, American imperialism has a rendezvous with disaster. It cannot conquer the world. It cannot reimpose colonial shackles upon the masses of the Middle East. It will not find through the medium of war a viable solution to its internal maladies. Rather, the unforeseen difficulties and mounting resistance engendered by war will intensify all of the internal contradictions of American society.

Given that the Iraq war is not an aberration, and that war is the inevitable product of a world situation dominated by the increasing tensions between a globally integrated economy and the capitalist nation state system (WSWS Editorial Board, ibid.); and given the aforementioned decline of U.S. imperialism, the WSWS Editorial Board (2008a) concludes, during the Clinton/Obama nomination contest, as follows:

> Today, an effective struggle against the war cannot be waged based on protests and appeals to the existing two-party system, or on yet another attempt to place greater power in the hands of the Democrats by putting Clinton or Obama in the White House and giving the party a larger majority in the Senate. What is required is a rejection of imperialism itself. Ending the wars in Iraq and Afghanistan and defeating the already well-advanced plans for further and even bloodier wars in Iran and elsewhere is possible only through the fight to mobilize the working class against the capitalist system that is the source of war.

In a secret videoconference in November, 2007, the puppet regime of Nouri al-Maliki and the Bush administration signed an agreement, ironically entitled 'Declaration of Principles for a Long-Term Relationship of Cooperation and Friendship' (Walsh, 2008). This contained plans for the establishment of permanent American military bases and gave preferential treatment for U.S. energy conglomerates and investors to exploit Iraqi oil reserves (Walsh, 2008). More recently, the U.S. military announced plans to keep at least 140,000 troops in Iraq indefinitely (WSWS, 2008b).

In addition, the United States is demanding that Iraq give it the authority to establish fifty permanent military bases throughout the country, together with other sweeping powers, such as ceding control of its airspace to American forces (Hassan Al-Sunaid, Iraqi member of parliament, cited in Van Auken, 2008c). This would allow U.S. forces to launch military operations without any prior consultation or permission (Al-Sunaid, cited in ibid.). As Van Auken (2008c) argues this 'would extend the present U.S. military occupation indefinitely and formalize the country's status as an American semi-colony'. According to Patrick Cockburn in the UK newspaper, *The Independent*, this would allow U.S. troops to 'arrest Iraqis and enjoy immunity from Iraqi law' (cited in Van Auken, 2008c). U.S. forces will be able to arrest Iraqis and imprison them indefinitely without charges, according to Cockburn (cited in ibid.). Tens of thousands remain in U.S. custody (Van Auken, 2008c). Meanwhile, citing 'senior Iraqi military sources' the *Gulf News* reported that the U.S.-Iraqi security agreement proposed by Washington would also include 'the right for the United States to strike, from within Iraqi territory, any country it considers a threat to its national security' (cited in Van Auken, 2008c), thus cementing 'the original aim of the illegal U.S. invasion: U.S. hegemony over the oil-rich Persian Gulf' (Van Auken, 2008c).

I do not know whether or not the Bush regime knows the work of Niall Ferguson, but it seems that his sentiments about the need for a permanent presence, and the relative unimportance of elections in the face of the loss of U.S. imperial power have been heeded. All this is not to do with 'white supremacy'. Rather, it relates directly to U.S. imperialist hegemony, itself connected to neoliberal racialized capitalism, hegemony and oil. To this effect, an Oil Law, drafted in secret and in consultation with International Oil Companies and the U.S. and UK governments, is on the verge of being passed. The Law will allow for the privatization of Iraqi oil for up to 30 years and permit regions to pass their own laws and sign their own contracts with oil companies (Hands Off Iraqi Oil, 2008). Oil production in Iraq is at its highest level since the 2003 invasion (Weaver, 2008a). According to Matthew Weaver (2008a), the Iraqi government wants to increase production by 20 percent, as the country has an estimated 115 billion barrels of crude reserves.

It is worth noting here that, in order to forestall any opposition from workers to these processes and to keep profits high, the legislation enacted by Saddam Hussein in 1987, which banned trade unions in the public sector and public enterprises (80% of all workers), is still in effect. This was enforced by Paul Bremer's post-invasion Occupation Authority and then by all subsequent Iraqi administrations (General Union of Oil Employees in Basra, 2008).

In this chapter, I have addressed some key issues related to neoliberal global capitalism, neoliberalism, global environmental destruction and twenty-first century imperialism, concluding with some comments on the occupation of Iraq five years on. In the next chapter, I will look at some common objections to Marxism, and attempt to answer them from a Marxist perspective. I will then consider the possibilities for twenty-first century socialism, examining developments in the Bolivarian Republic of Venezuela. I will conclude with a discussion of antiracism in practice.

Appendix

Here is the Interrogation Log of Detainee 063 at Camp X-Ray in Guantanamo Bay:

Day 25, December 17, 2002
0120: Control shows detainee photos from a fitness magazine of scantily-clad women.
1400: Detainee was shown 9/11 tribute videos.

2100: Detainee did not appreciate being called a homosexual. He also appeared annoyed by the issue of his mother and sister as examples of prostitutes and whores.

Day 27

1100: Happy Mohammed mask placed on detainee and he was yelled at when he tried to speak.

2320: He attempts to resist female contact. He attempts to pray as she spoke in his ear about his continuous lies...

1940: Sgt M had shown detainee a picture of Mecca, there were thousands of Muslims congregated...Detainee broke down and cried.

Day 28

1115: Told detainee that a dog is held in higher esteem because dogs know right from wrong. Began teaching the detainee lessons such as stay, come, and bark...Detainee very agitated.

1300: A towel was placed on the detainee's head like a burka, with his face exposed, and the interrogator proceeded to give him dance lessons.

2200: The detainee was strip-searched. After five minutes of nudity, the detainee ceased to resist...

Day 29

2103:...I was forehead to forehead with the detainee and he stated that he would rather be beaten with electrical wire than have me constantly in his personal space...

Day 31

0100:...lead (interrogator) hung pictures of swimsuit models around his neck.

Day 32

1145: Detainee refused water so control poured a little on his head.

2100: Detainee seems to be on the verge of breaking.

Day 33

0300: Detainee started falling asleep so interrogator had detainee stand up for 30 minutes. Detainee was subjected to white noise (music) waiting for his IVs to be completed.

Day 50

0230: Source received haircut...Detainee stated he would talk about anything if his beard was left alone. Beard was shaven...detainee began to cry when talking.

Excerpted from a daily log. (Sands, 2008, pp. 21–23)

Chapter 7

Marxism and Twenty-first-Century Socialism

Throughout this book, I have constantly invoked Marxism as being more conducive to understanding racism and its relationship to capitalist society than CRT; and have defended Marxism against CRT critiques of it. In order to both substantiate my defense of, and indeed, exaltation of the modern Marxist project; in order specifically to argue CRT attempts to render it passé, no longer relevant, racist and oppressive, it is incumbent on me to justify the overall strengths of Marxism as a worldview. The most effective way to do this, I would argue, is to address some common objections to Marxism, and to attempt to answer them from a Marxist perspective (this will also provide a backdrop and a base, as well as some political energy, for my discussion of classroom practice in the next chapter). After doing this, in order to further make the case that Marxism is *not* a spent force and is relevant to the twenty-first century, I then look, as a case study, at ongoing developments in the Bolivarian Republic of Venezuela, focusing on the notable social democratic changes designed to improve the lives of the Venezuelan working class. These include a number of very impressive reforms to the education system, which serve as a beacon of enlightenment when compared to the educational practices in the United States and the United Kingdom that antiracist educators are challenging (see chapter 4 of this volume). I go on to discuss the potential in Venezuela for twenty-first-century socialism. To counter CRT claims of an incompatibility between Marxism and antiracism, I conclude the chapter with a discussion of the way in which President Hugo Chávez and others are championing indigenous and Afro-Venezuelan rights as part of what they perceive to be the transition to socialism (given the prominence I give to the voice of Hugo Chávez, a Venezuelan with both indigenous and African roots, hopefully my analysis in this part of the chapter might strike a chord with Critical Race Theorists). As a result of developments in antiracist thinking, *including* CRT, it is my view that a 'color-neutral' socialism, like a socialism that does not take account of gender (thanks to contributions by feminists and their supporters) is now permanently off the progressive agenda.

According to Marxist theory, the anarchy of capitalist production creates the material conditions for the proletarian revolution (the transformation of the socialized means of production from the hands of the bourgeoisie into public property) (Engels, 1892 [1977], p. 428). While, for Marxists, proletarian revolution is not *inevitable*, it is always on the cards. As Tom Hickey (2006) explains, capitalism has an inbuilt tendency to generate conflict, and is thus *permanently* vulnerable to challenge from the working class. As he puts it:

> The objective interests of the bourgeoisie and the proletariat are incompatible, and therefore generate not a tendency to permanent hostility and open warfare but a permanent tendency toward them. The system is thus prone to economic class conflict, and, given the cyclical instability of its economy, subject to periodic political and economic crises. It is at these moments that the possibility exists for social revolution. (Hickey, 2006, p. 192)

At times, the ruling class may for forced into a settlement. For example, to take the case of the United Kingdom, during the postwar Labour Government of Clement Attlee (1945–1951), in order to foreclose the possibility of social revolution at the time, the ruling class was prepared to allow a series of major reforms. Thus the period saw sweeping legislation that began with the Beveridge Report of 1942 (Beveridge, 1942). Attlee's government created a national insurance system, the National Health Service and embarked on a massive program of nationalization. Given the political volatility of the period after World War II, these measures, which created the British Welfare State, may be seen as a compromise between capital and labor (Cole, 2008d, p. 1).[1]

Another example is the General Strike of 1926. This ended a sixteen year period of intense industrial militancy by the British working class, which had commenced with the Great Unrest of 1910–1914 and which continued through the engineering struggles during the First World War (Davidson, 2006). This reached its climax in the mass strikes of 1919, which, according to Davidson, 2006, was the point of maximum danger for the British ruling class, and its European counterparts more generally. The actual events of 1926 were less threatening, in part because they were held back by the trade union leaders (Davidson, 2006).

Outside of these political and economic crises, everything under capitalism has a certain, at times hidden, at times transparent, class-based, racialized and gendered logic of inevitability and insurmountability, prompted in part by the success of the state apparatuses in interpellating subjects—this is how things are or even should be, and there's nothing we can or even should do (see chapter 2 of this volume).[2] The ruling class's success at keeping Marxism off the agenda, most notably in the United States, and significantly in the United Kingdom since Thatcherism and its aftermath (see Cole, 2008g) is not logical (indeed, given that Marxism is in the interests of the working class, it is, in fact, *illogical*). However, as Stuart Hall (1978) once remarked, ideologies don't work by logic—they have logics of their own. Thus:

> we act and respond to ideology as if we were the originators of the ideas and values within it. In other words, when *The Sun* or *The Daily Mail* [3] speaks of what 'the public' 'wants', 'needs', 'is fed up with', 'has had enough of' this strikes a chord with all the other organs of ruling-class ideology—the rest of the media, the various apparatuses of the state. Because we are largely trapped with one view of the world . . .—it all makes sense to us. (Cole, 1986b, p. 131)

It is the role of Marxism and Marxists to transcend these ruling class interpellations, to provide an alternative vision, an insistence that another world is possible. I will now address some of the common objections to Marxism, themselves by and large the result of successful interpellation, and attempt to respond to them.

Common Objections to Marxism and a Marxist Response[4]

How Is the Marxist Vision of Socialism Different from Capitalism and Why Is It Better?

Marxists do not have a blueprint for the future (see Rikowski, 2004, pp. 559–560; see also Gibson and Rikowski, 2004 and Cole, 2008d, pp. 80–81). However, there are certain features, which would distinguish world socialism from world capitalism. What follows are just a few examples.

Bowles and Gintis (1976, p. 54) argue that whereas in capitalist societies, the political system is 'formally democratic', capitalist economies are 'formally totalitarian', involving:

> the minimal participation in decision-making by the majority (the workers); protecting a single minority (capitalists and managers) against the wills of a majority; and subjecting the majority to the maximal influence of this single unrepresentative minority.

Under socialism, this would be reversed. The workers would own and control the means of production and would encourage maximal participation in decision-making.

Public services would be brought under state control and democratically run by the respective workforces. There would be universal free health care for all, incorporating the latest medical advances. There would be no need for private health. There would universal free comprehensive education for all and no need for private schooling. There would be free comprehensive leisure facilities for all, with no fee for health clubs, concerts etc. There would be free housing and employment for all. There would be full rights for women, for the LGBT (lesbian, gay, bisexual and transgender) communities, for all members of minority ethnic groups and for disabled people. There would be full freedom of religion.[5] There would be no child pornography. There would be no war, no hunger and no poverty.

The substantive issues in the above paragraph are not made up, but form the platform for (the British) Respect the Unity Coalition (undated) which campaigns on a broadly current-day Marxist platform.

Bowles and Gintis (1976, p. 266) capture the essence of socialism as follows:

> Socialism is not an event; it is a process. Socialism is a system of economic and political democracy in which individuals have the right and the obligation to structure their work lives through direct participatory control. [Socialism entails] cooperative, democratic, equal, and participatory human relationships; for cultural, emotional and sensual fulfilment.

Marxism Is Contrary to Human Nature Because We Are All Basically Selfish and Greedy and Competitive

For Marxists, there is no such thing as 'human nature'. Marxists believe that our individual natures are not ahistorical givens, but products of the circumstances into which we are socialized, and of the society or societies in which we live or have lived (including crucially *the social class position* we occupy therein). While it is true that babies and infants, for example, may act selfishly in order to survive, as human beings grow up they are strongly influenced by the norms and values that are predominant in the society in which they live. Thus in societies which encourage selfishness, greed and competitiveness (Thatcherism is a perfect example) people will tend to act in self-centered ways, whereas in societies which discourage these values and promote communal values (Cuba is a good example) people will tend to act in ways that consider the collective as well as their own selves, the international, as well as the national and local.[6] As Marx (1845) put it, '[l]ife is not determined by consciousness, but consciousness by life'. Unlike animals, we have the ability to choose our actions, and change the way we live, and the way we respond to others. Hence, in capitalist society, the working class *is* capable of transcending false consciousness and becoming, 'class for itself' (Marx, 1847 [1995]), as well as 'a class in itself' (Bukharin, 1922 cited in Mandel, 1970 [2008]), that is to say, pursuing interests which can ultimately lead to a just society. Socialism does not require as a precondition that we are all altruistic and selfless; rather, as Bowles and Gintis (1976, p. 267) argue, the social and economic conditions of socialism will facilitate the development of such human capacities.

Some People Are Naturally Lazy and Won't Work

Unlike utopian socialist Henri de Saint-Simon, who believed that we are 'lazy by nature' (Cole, 2008d, pp. 15–17),[7] Marxists would argue that laziness, like other aspects of our 'nature' is most likely acquired through socialization too, but even if it is not, we can still choose to overcome our laziness. In a socialist world, there are sound reasons to work, in order to create cooperative wealth. Whereas in capitalist societies, a surplus is extracted from the values workers produce, and hived off by the capitalists to create profits, under socialism everything we create is for the benefit of humankind as a whole, including us as individuals. Thus the only incentive for most workers under capitalism: more wages (an incentive which is totally understandable, and indeed encouraged by Marxists, because it ameliorates workers' lives and lessens the amount of surplus for capitalists) is replaced by a much more worthwhile incentive, the common good.

Why Shouldn't Those Who Have Worked Hard Get More Benefits in Life?

Again this viewpoint is a product of capitalist society, based on selfishness. If everything is shared, as in a socialist world, we all benefit by working hard. No one needs to go short of anything that they need for a good life. Of course, in capitalist societies, needs are created by advertisers for capitalists, and many of these (excessive amounts of clothes, or living accommodation beyond our personal requirements) we do not really *need*. Indeed, excess possessions in a capitalist world where most of the population has nothing or next to nothing is obscene.

Moreover, while people are starving and there are food riots breaking out across the world, the fact is, given the world's total resources, there is no world food shortage overall, and there is more than enough food produced to supply everyone with a decent diet (Molyneux, 2008, p. 13). The grotesque nature of capitalism is revealed *par excellence* at the time of writing (Summer 2008). UK prime minister Gordon Brown's response to the horrendous rise in the cost of living in the United Kingdom was that people—working class people presumably because this would not matter to the rich—should cut the amount of food they throw away. He followed this up just hours later by attending a G8 summit in Japan (July 2008) where during the course of one day, he gorged on 14 courses of 24 different dishes and at least 5 different wines. Part of the agenda of this G8 meeting is the global issue of the immediate dangers of the soaring prices of food!

Marxism Can't Work Because It Always Leads to Totalitarianism

Marxists have learnt from Stalinism, which was, in many ways, the antithesis of Marxists' notions of democratic socialism. While not in any way condoning it, part of the reason for Stalinist totalitarianism is that socialism was attempted in one country, whereas Marx, and a number of Marxists at the time (notably Trotsky) believed that, for it to work, it must be international. This meant that the Soviet Union, being isolated, concentrated on accumulation rather than consumption. Alone in a sea of capitalist states, the economy was geared to competing economically and militarily with the rest of the world, with workers' rights taking a back seat. I am not claiming that this direction for the Soviet Union was *inevitable* and there are no inherent reasons why these mistakes should be made again. To succeed, socialism needs to be democratic. Indeed, as Jonathan Maunder (2006, p. 13) reminds us, whereas previous exploited classes, such as the peasantry could rise up, seize lands and divide them up among themselves, workers cannot, for example, divide a factory, hospital or supermarket. Thus if workers do seize control of such institutions, they can only run them collectively. As Maunder (2006, p. 13) concludes: '[t]heir struggles have a democratic logic that can lay the basis for a different way of running society'.

Genuinely *democratic* socialism, where elected leaders are permanently subject to recall democratically by those who have elected them, is the best way to safeguard against totalitarianism (this concept, a central plank of democratic socialism, is in fact enshrined in the 1999 constitution of the Bolivarian Republic of Venezuela in the form of a Recall Referendum. This means that Venezuelan voters have the right to remove their president from office before the expiration of the presidential term).

While, following Bowles and Gintis, I noted above that capitalist political systems are *formally* democratic, bourgeois democracy, for example in Britain and the United States, in effect amounts to a form of totalitarianism. In these countries, citizens can vote every five years, having in reality a choice (in the sense of who will actually be able to form a government) of two main totally pro-capitalist parties, who then go on to exercise power in the interests of neoliberal global capitalism and imperialism with little or no regard for the interests of those workers who elected them. There are, of course, some restraints on what they can get away with (minimum wage and European human rights legislation in Britain, for example), and importantly, the balance of class forces and the strength of working class resistance (e.g., Hill, 2009a).

Someone Will Always Want to Be 'Boss' and There Will Always Be Natural 'Leaders' and 'Followers'

As argued above, Marxists believe in true democracy. If a given individual in socialist society wants to exploit others, s/he will need to be controlled democratically and subject to permanent recall. Under capitalism, if people feel they are 'born to be followers' rather than leaders, this is most like to be due to their social class position in any given society and to their socialization (see above). Under socialism, there will be more chance for all to take roles of responsibility if they want. Under capitalism, see certain people are educated for leadership positions in the society, while others are schooled to be exploited members of the working class (Bowles and Gintis, 1976).

It Is Impossible to Plan Centrally in Such a Hugely Diverse and Complex World

In a socialist world, local, national and international needs will need to be coordinated fairly and efficiently. Given modern technology, this is easier now than ever before, and will become more and more so, as technology continues to develop. Under capitalism, technology is harnessed to the creation of greater and greater surplus value and profit. In a socialist world, technology would be under the control of the people for the benefit of the people as a whole; for universal human need rather than global corporate profit. Cuba is a good example.

Someone Has to Do the Drudge Jobs, and How Could that Be Sorted Out in a Socialist World

Technology already has the potential to eliminate most of the most boring and/or unpleasant jobs. Some of those that remain could be done on a voluntary rota basis, so that no one would have to do drudge jobs for longer than a very brief period (utopian socialist Charles Fourier had a similar idea—see Cole, 2008d, pp. 17–20). Voluntary work under capitalism in the public sector abounds, and there is every reason to assume that such work would flourish much more under socialism.

Socialism Means a Lower Standard of Life for All

World socialism will only lower the standard of life for the Ruling Classes. There will not, for sure, be the massive disparities of wealth apparent in our present capitalist world. There will, of course, be no billionaires and no need for a (parasitic) monarchy. If the wealth of the world is shared, then there will be a good standard of life for all, since all reasonable needs will be met, including enough food (as noted above by Molyneux (2008, p. 13) enough already exists). To paraphrase Marx, 1875, the principle will be from each according to his or her ability, to each according to his or her needs.

Socialism Will Be Dull, Dreary, and Uniform and We Will All Have Less Choice

This is a popular misconception related to the experiences of life in former Stalinist states such as the USSR, and the former states of Eastern Europe. Life under socialism should be exciting, challenging and globally diverse, as different countries develop socialism to suit their own circumstances, but with a common goal. The intensively creative (world) advertising industry (now in private hands), under public control could be used for the common good, for example to increase awareness of the availability of free goods and services (health promotion, universal life-long education, public transport and so on). We do not need the excessive branded products common in capitalist societies, and created by different capitalist firms to increase profits. A cursory glance at the Web site of one well-known supermarket in Britain revealed a total of over sixty different butters/margarines. It is not necessary for western consumers to have this degree of 'choice' when most of 'the developing world' eats its bread without spreads. Moreover, in many cases the ingredients in the vast array of products will be very similar, while the huge amount of unnecessary plastic packaging clearly adversely affects the environment (see pp. 99–100 of this volume, see also, Cole, 2008d, Chapter 7).

A Social Revolution Will Necessarily Involve Violence and Death on a Massive Scale

It is, in fact, capitalism that has created and continues to promote death and violence and terror on a global scale. Inequalities in wealth and quality of life cause death and disease in capitalist countries themselves, and the capitalist west's underdevelopment of most of the rest of the world and the aforementioned massive disparity in wealth and health has dire consequences (Hill and Kumar, 2009; Hill and Rosskam, 2009). In addition, imperialist conquest historically and contemporaneously unleashes death, terror and destruction on a colossal scale. Stalinism, and other atrocities, committed *in the name of,* but not in the spirit of socialism, also shares this guilt, but as argued above, there is no inherent reason why the historical perversities of Stalinism need to be repeated. As for the violence entailed in future social revolution is concerned, this is, of course, an unknown. However, as argued in Cole, 2008d, pp. 78–79, socialism is a majoritarian process not an imposed event which is not *dependent* on violence. It is, of course, inconceivable that a world social revolution would involve no violence, not least because of the resistance of the dominant capitalist class. However, there are no reasons for violence to be a strategic weapon. Anyone who has ever attended a mass socialist gathering, e.g., *Marxism 2008* in Britain (http://www.marxismfestival.org.uk/), can attest to the fact that violence is not, in any way, an organizing tool of the socialist movement. Mass violence is the province of world capitalism. Moreover, Marxists oppose terrorism unreservedly. Terrorism is reactionary, in that it diverts attention away from the class struggle. It militates against what Leon Trotsky has described as self-organization and self-education. Trotsky favored a different resolution to the revenge desired by many who subscribe to terrorism. As he put it:

> The more 'effective' the terrorist acts, the greater their impact, the more they reduce the interest of the masses in self-organisation and self-education...To learn to see all the crimes against humanity, all the indignities to which the human body and spirit are subjected, as the twisted outgrowths and expressions of the existing social system, in order to direct all our energies into a collective struggle against this system—that is the direction in which the burning desire for revenge can find its highest moral satisfaction. (Trotsky, 1909)

The Working Class Won't Create the Revolution Because They Are Reactionary

It is a fundamental tenet of Marxism that the working class are the agents of social revolution, and that the working class, as noted above, needs to become a 'class for itself' in addition to being a 'class in itself' (Marx, 1847 [1995]). It is unfortunately the case that major parts of the world are a long way off such a scenario at the present conjuncture. It is also the case that successful interpellation and related false consciousness hampers the development of class consciousness and the move towards the overthrow of capitalism. Britain is one example where the Ruling Class has been particularly successful in interpellating the working class (see Cole, 2008g, 2008h for a discussion).

Elsewhere, however, there are examples of burgeoning class consciousness, witnessed for example by the growth of Left parties (see below) in Europe and by developments across South America, notably the Bolivarian Republic of Venezuela (see below) and in Bolivia. It is to be hoped that, as neoliberal global imperial capitalism continues to reveal and expose its essential ruthlessness and contempt for those who make its profits, class consciousness will increase and that the working class will one day be in a position to overthrow (world) capitalism and to replace it with (world) democratic socialism. Perhaps it should be pointed out here that Marxists do not idolize or deify the working class; it is rather that the structural location in capitalist societies of the working class, so that, once it has become 'a class in itself' makes it the agent for change. Moreover the very act of social revolution and the creation of socialism mean the end of the very existence of the working class as a social class. As Marx and Engels (1845) [1975] put it:

> When socialist writers ascribe this world-historic role to the proletariat, it is not at all...because they regard the proletarians as *gods*. Rather the contrary...[The proletariat] cannot emancipate itself without abolishing the conditions of its own life. It cannot abolish the conditions of its own life without abolishing all the inhuman conditions of society today which are summed up in its own situation.

Marxists Just Wait for the Revolution Rather than Address the Issues of the Here and Now

This is manifestly not the case. As I have argued earlier, Marxists fight constantly for change and reform which benefit the working class in the short-run under capitalism (for example, Marxists are centrally involved with work in trade unions agitating for better wages) with a vision of socialist transformation in the longer term (increasing class consciousness in the unions is part of this process). As Marx and Engels, 1847 [1977a], p. 62 put it:

> The Communists fight for the attainment of the immediate aims, for the enforcement of the momentary interests of the working class; but in the

movement of the present, they also represent and take care of the future of the movement.[8]

The *Respect* Coalition has this dual aim, combining reform (bullet point 2: 'the fight against') with a revolutionary vision (bullet point 2: 'the ultimate abolition' and bullet points 1, 3 and 4):

- The organization of society in the most open, democratic, participative, and accountable way practicable based on common ownership and democratic control.
- The fight against, and ultimate abolition of racism, sexism, and all other forms of discrimination on the grounds of religion, disability, age or sexual identity. Defend a woman's right to choose.
- The abolition of all forms of economic exploitation and social oppression.
- The promotion of peace and a system of global and national justice that provides protection from tyranny, prejudice and the abuse of power. (Respect, the Unity Coalition, undated, p. 3)

However, there is no illusion that getting into power in the local and national capitalist state will create socialism, but it does provide a space to spread the message. As Respect's MP George Galloway put it:

We don't believe that the world can be changed in town halls and in parliament, but we believe that town halls and parliaments can be used to build a mass movement of people that *will* change things in this country for the better. (Galloway, 2006)

The choice is not between life in the neoliberal global capitalist world or a return to Stalinism, but between the anarchic chaos of capitalism and genuine world wide democratic socialism. There is a burgeoning recognition that this is the case from the mass global movements against globalization and in the growing anti-neoliberal politics throughout Latin America, from the Bolivarian Republic of Venezuela to Bolivia, from Argentina to Brazil. There is also growing support for some Marxist parties in Europe, such as the Die Linke in Germany, the Portuguese Bloco Esquerda the Dutch Socialist Party, and the Red-Green Alliance in Denmark.

Marxism Is a Nice Idea, But It Will Never Happen (For Some of the Reasons Headlined Aabove)

Bringing Marxism to the forefront is not an easy task. Capitalism is self-evidently a resilient and very adaptable world force and, as I have argued throughout this volume, interpellation has been very successful. However, Marx argued that society has gone through a number of different stages in its history: primitive communism; slavery; feudalism, capitalism. It is highly likely that in each era, a different way of living was considered 'impossible' by most of those living in that era. However, each era gave birth, in a dialectical process, to another. Thus, though it may be extremely difficult to imagine a world based on socialist principles, such a world *is* possible if that is what the majority of the world's citizens come to desire and have the will to create. Marxists need to address the obstacles full on. As Callinicos (2000, p. 122) has argued, we must break through the 'bizarre ideological mechanism, [in which] *every* conceivable alternative to the market has been discredited by the collapse of Stalinism' whereby the fetishization of life makes capitalism seem natural and therefore unalterable and where the market mechanism 'has been hypostatized into a natural force unresponsive to human wishes' (p. 125).[9] Capital presents itself 'determining the future as surely as the laws of nature make tides rise to lift boats (McMurtry, 2000, p. 2), 'as if it has now replaced the natural environment. It announces itself through its business leaders and politicians as coterminous with freedom, and indispensable to democracy such that any attack on capitalism as exploitative or hypocritical becomes an attack on world freedom and democracy itself' (McLaren, 2000, p. 32).[10] However, the biggest impediment to social revolution is not capital's resistance, but its success in heralding the continuation of capitalism as being the only option. As Callinicos puts it, despite the inevitable intense resistance from capital, the 'greatest obstacle to change is not...the revolt it would evoke from the privileged, but the belief that it is impossible' (2000, p. 128). Given the hegemony of world capitalism, whose very *leitmotif* is to stifle and redirect class consciousness, and given the aforementioned reactionary nature of certain sections of the working class, restoring this consciousness is a tortuous, but not impossible task. Callinicos again:

> Challenging this climate requires courage, imagination and willpower inspired by the injustice that surrounds us. Beneath the surface of our supposedly contented societies, these qualities are present in abundance. Once mobilized, they can turn the world upside down. (2000, p. 129)

As we hurtle into the twenty-first century, we have some important decisions to make. Whatever the twenty-first century has to offer, the choices will need to be debated. The Hillcole Group expressed our educational choices as follows:

> Each person and group should experience education as contributing to their own self-advancement, but at the same time our education should ensure that at least part of everyone's life activity is also designed to assist in securing the future of the planet we inherit—set in the context of a sustainable and equitable society. Democracy is not possible unless there is a free debate about all the alternatives for running our social and economic system...All societies [are] struggling with the same issues in the 21st century. We can prepare by being better armed with war machinery or more competitive international monopolies...Or we can wipe out poverty...altogether. We can decide to approach the future by consciously putting our investment into a massive drive to encourage participation from everyone at every stage in life through training and education that will increase productive, social, cultural and environmental development in ways we have not yet begun to contemplate. (Hillcole 1997, pp. 94–95)

While the open-endedness of the phrase, 'in ways we have not yet begun to contemplate' will appeal to poststructuralists and postmodernists, for whom the future is an open book, this is most definitely not the political position of the Hillcole Group. Whereas, for poststructuralists and postmodernists, all we have is endlessly deconstruction without having *strategies* for change (see Cole, 2008d, Chapter 5), for Marxists, the phrase is tied firmly to an open but *socialist* agenda.

For transmodernists such as David Geoffrey Smith, the way forward is to 'desacralize capitalism' and to move towards a 'rethought liberal democracy', while for Enrique Dussel, the answer lies in an 'ex nihilo utopia' (a utopia from nothing) (for a critique of these positions, see Cole, 2008d, pp. 75–84). For Critical Race Theorists, there are nonspecific notions of 'ending oppression' (see the Conclusion to this volume). These suggestions are no doubt well intentioned, but they are idealistic in the current historical conjuncture. Like the views of the utopian socialists (Engels, 1892), neither Smith's nor Dussel's ideas nor the vagaries of CRT engage with the nature of the contradictions within capitalism, the dialectic, and with the working class consciousness needed for revolutionary change.

An equitable, fair and just world can be foreseen neither through postmodernism/post-structuralism, nor through the more enlightened and progressive ideas of transmodernism and CRT. For Marxists, as global neoliberal capitalism and imperial hegemony tightens its grip on all our lives, the choice, to paraphrase Rosa Luxembourg (1916), is quite simple: that choice is between barbarism—'the unthinkable'—or democratic socialism.

Ok, Show Me Where Marxism Works in Practice

Even if all of the above questions are answered convincingly, Marxists are inevitably asked, 'Ok, show me where Marxism works in practice?' I have lost count of the number of times in a lifetime working in education that I have been asked that question. Since I first visited Cuba a decade or so ago, and up to my trip to Venezuela where I worked briefly for the Bolivarian University of Venezuela in 2006 (see Scott, 2006, p. 14), I tended to reply on the lines of, 'well, I know it's not perfect, but the case of Cuba is in many respects a good example'. However, I am now able to commend developments in Venezuela with far fewer reservations. Elsewhere (Cole, 2008g, 2008h) I have discussed at length the Twenty-first-Century Socialism advocated by President Hugo Chávez in Venezuela. Here I will present a summary.

The Bolivarian Revolution

In Venezuela, neoliberal capitalism is not seen as 'inevitable', nor indeed is capitalism itself. President Hugo Rafael Chávez Frías[11] is against neoliberalism and imperialism. As he remarked in 2003:

In Venezuela, we are developing a model of struggle against neoliberalism and imperialism. For this reason, we find we have millions of friends in this world, although we also have many enemies. (cited in Contreras Baspineiro, 2003)

Chávez talks about globalizing socialism, rather than global capitalism:

> Faced with the outrageous excesses of the powerful, our only alternative is to unite…That's why I call upon all of you to globalize the revolution, to globalize the struggle for…freedom and equality. (cited in Contreras Baspineiro, 2003)

In January, 2007, the utilities sector and the country's largest telecoms company were nationalized, and in May, 2007, Chávez took control of four major oil projects worth $30 billion (PDVSA, the state oil company has held majority control since 2005). Also in May, 2007, Venezuela pulled out of the IMF and the World Bank.

Venezuela's only steel producer, SIDOR (Siderúrgica del Orinoco), has, at the time of writing, been renationalized following an announcement by President Chávez in April 2008. Chávez has insisted that the workers must set up workers' councils, and that SUTISS, the trade union representing workers at SIDOR, exercise a policy of workers' control and management (J. Martin, 2008). Juan Valor, the Press Secretary of SUTISS, stated that, with the renationalization, 10,000 workers had been 'liberated', noted that 'we know that our commitment with Venezuela, with the people and with President Chavez is even greater'. Oil, steel and cement have all been declared strategic industries, and according to the government's economic and political program, must be brought under public control (Venezuela Information Centre Update, April 24, 2008, info@vicuk.org).

In July, 2008, Chávez announced that the Venezuelan Government will nationalize the Bank of Venezuela, the country's third largest bank. Complementing the private management for achieving a 'high level of efficiency' in the bank, he made it clear that his administration plans to shift the bank's priorities from profits to social investments. As he put it, '[t]he profit will not be of one group, but to invest in socialist development' (cited in Venezuelanalyisis.com, 2008).[12]

The Misiones

Venezuela, of course, has vast reserves of oil, and will have for another hundred years, according to Chávez (Campbell, 2008, p. 58).[13] The wealth that this generates is being used to fund a vast program of reform. Central to the Bolivarian Revolution, then, are the 'Misiones'—a series of social justice, social welfare, antipoverty, and educational programs implemented under the administration of the Chávez government. The Misiones use volunteers to teach reading, writing and basic mathematics to those Venezuelans who have not completed their elementary-level education. In addition, they provide ongoing basic education courses to those who have not completed such education, and remedial high school-level classes to millions of Venezuelans who were forced to drop out from high school.

The Bolivarian University of Venezuela (UBV) is particularly significant in the struggle for social justice. Previously the luxurious offices of oil oligarchs, UBV has opened its doors to thousands of students. This program's goal is to boost institutional synergy and community participation in order to guarantee and provide access to higher education to all high school students.[14] UBV had its first 1078 graduates (70% of which are women) in 2006 and has to be seen in the context of the established university system in Venezuela. Like many others in Latin America, has traditionally primarily served a limited, better-off section of the population. Access for the poor majority has been extremely restricted, partly because of the financial costs of university study, but also because of a deeply entrenched system of corruption and patronage governing entry procedures etc. Since 1998, the government has raised the number of university students in the country from 366,000 to 1,200,000. This is a genuine widening participation initiative.[15]

Chávez has announced that he aims to create 38 new universities in the current phase of the 'Bolivarian Revolution'. The state universities go out to the people in the barrios, as well as the people coming to them, with the government aiming for more than 190 satellite classrooms throughout Venezuela by 2009.

The Misiones are not confined to education. Other Misiones are concerned with health, providing a free service to the poor by giving access to health care assistance to 60 percent of the excluded population through the construction of 8,000 Popular Medical Centers: providing a doctor for every 250 families (1,200 people), increasing the life expectancy of the population, and contributing to a good standard of life for all.[16]

Yet other Misiones aim at assisting the sport skills of students, senior citizens, pregnant women, people with disabilities, and anyone wishing to improve their standard of life and health, and include high performance sport, and vocational training for work. There is also a mission that sells food and other essential products like medicines at affordable prices, along with a massive program of soup kitchens (British Venezuela Solidarity Campaign, 2006).

Crucially, there are Misiones to restore human rights to numerous indigenous communities, and to hand over land titles to farmers in order to guarantee food for the poor and to foster a socialized economy and endogenous development (ibid.).

Negra Hipólita Mission, one of the newest created by the National Government, was launched on January 14, 2006, in order to fight poverty, misery and social exclusion; a new stage in the struggle against inequality.

At the time of writing (July 2008) The Venezuelan government has announced that it is to build 50,000 houses for poor families in six urban complexes. An additional 5,000 apartments are under construction as part of a project to build 18,000 similar dwellings. The apartments are being built under a government program called Mission Villanueva (Venezuela Information Centre, July 3, 2008).

The Bolivarian Republic of Venezuela is committed to set Venezuela free from misery (ibid.). For a full description of the Misiones see the see British Venezuela Solidarity Campaign (2006) viewable online.

The Misiones are, of course, classic examples of social democracy, somewhat akin to the policies and practice of the postwar Labour Governments in Britain. What distinguishes the Bolivarian Revolution, however, is that these reforms are seen both by the Chávez Government and by large sections of the Venezuelan working class as a step on the road to true socialist revolution, since for Chávez '[t]he hurricane of revolution has begun, and it will never again be calmed' (cited in Contreras Baspineiro, 2003). Elsewhere, Chávez asserted: 'I am convinced, and I think that this conviction will be for the rest of my life, that the path to a new, better and possible world, is not capitalism, the path is socialism, that is the path: socialism, socialism' (Lee, 2005).

Marxism and the Venezuelan State

At this point it is useful to return to Althusser and to the Marxist theory of the State. In classical Marxist theory, the capitalist state must be overthrown rather than reformed. As Althusser (1971: 142) put it:

> the proletariat must seize State power in order to destroy the existing bourgeois State apparatus and, in a first phase, replace it with a quite different proletarian, State apparatus, then in a later phases set in motion...the end of State power, the end of every State apparatus.

However, Althusser's analysis did not extend to the possible existence of states which advocate their own destruction. As Chávez proclaimed at the World Social Forum in 2005:

> It is impossible, within the framework of the capitalist system to solve the grave problems of poverty of the majority of the world's population. We must transcend capitalism. But we cannot resort to state capitalism, which would be the same perversion of the Soviet Union. We must reclaim socialism as a thesis, a project and a path, but a new type of socialism, a humanist one, which puts humans, and not machines or the state ahead of everything. That's the debate we must promote around the world (my emphasis). (cited in Curran, 2007)

On January 8, 2007, Chávez created 'communal councils' and has referred to 'the revolutionary explosion of communal power, of communal councils' (Socialist Outlook Editorial, 2007). This is a project for rebuilding or replacing the bourgeois administrative machinery of local and state governments with a network of communal councils, where the local populations meet to decide on local priorities and how to realize them (ibid.). 'With the communal councils', Chávez said, in perhaps his most clearly articulated intention to destroy the existing state:

> we have to go beyond the local. We have to begin creating...a kind of confederation, local, regional and national, of communal councils. We have to head

towards the creation of a communal state. And the old bourgeois state, which is still alive and kicking—this we have to progressively dismantle, at the same time as we build up the communal state, the socialist state, the Bolivarian state, a state that is capable of carrying through a revolution. (cited in ibid.)

'Almost all states', Chávez continued, 'have been born to prevent revolutions. So we have quite a task: to convert a counter-revolutionary state into a revolutionary state' (cited in Piper, 2007a: 8). The communal councils are intended to bring together 200 to 400 families to discuss and decide on local spending and development plans. Thirty thousand communal councils are intended, and provide, in the words of Roland Dennis, an historic opportunity to do away with the capitalist state (cited in Piper, 2007a).

If it is the case that genuinely supports socialist revolution from below, which will eventually overthrow the existing capitalist state of Venezuela, then, for Marxists, he must be seen as an ally. Whether he is or not, however, is less important than the fact that he is openly advocating and helping to create genuine socialist consciousness among the working class. I thus make no apologies for making Chávez's pronouncements a central feature of this chapter.

For example, swearing in the new ministers, in the wake of his landslide presidential election victory, late in 2006, Chávez declared that they will be in charge of pushing forward his government's project of implementing '21st century socialism' in Venezuela (Wilpert, 2007), which Chávez defines as 'fundamentally human, it is love, it is solidarity, and our Socialism is original, indigenous, Christian and Bolivarian' (cited in Hampton, 2006). More recently, Chávez advised all Venezuelans to read and study the writings of Leon Trotsky, and commented favorably on The Transitional Programme, which was written by Trotsky for the founding congress of the Fourth International in 1938 (J. Martin, 2007). Trotsky's pamphlet begins with a discussion of the objective prerequisites for a socialist revolution.

Trotsky's concept of 'the permanent revolution', Chávez went on, is an extremely important thesis (J. Martin, 2007). Chávez underlined Trotsky's idea about the necessity for conditions for socialism to be ripe and expressed his view that this is certainly the case in Venezuela (ibid.).

Chávez continued:

Trotsky points out something which is extremely important, and he says that [the conditions for proletarian revolution] are starting to rot, not because of the workers, but because of the leadership which did not see, which did not know, which was cowardly, which subordinated itself to the mandates of capitalism, of the great bourgeois democracies, the trade unions'. (cited in J. Martin, 2007)

For me, this statement is indicative of Chávez' belief in the importance of grass roots working class consciousness and action.

Jorge Martin agrees:

[s]ince Chávez started talking about socialism in January 2005, this has become a major subject of debate in all corners of Venezuela. Chávez's statement that

under capitalism there was no solution for the problems of the masses and that the road forward was socialism represented a major step forward in his political development. He had started trying to reform the system and to give the masses of the Venezuelan poor decent health and education services and land, and he had realised through his own experience and reading that this was not possible under capitalism. (ibid.)

Chávez has made clear that when he talks of building socialism, he is talking about doing it now, not in the long distant future (ibid.). In his comments about Trotsky he stressed the point:

> Well, here the conditions are given, I think that this thought or reflection of Trotsky is useful for the moment we are living through, here the conditions are given, in Venezuela and Latin America, I am not going to comment on Europe now, nor on Asia, there the reality is another, another rhythm, another dynamic, but in Latin America conditions are given, and in Venezuela this is a matter of course, to carry out a genuine revolution. (cited in ibid.)

Leading figures in some of the Bolivarian parties have refused to join in Chávez' new United Socialist Party, the United Socialist Party of Venezuela (PSUV) formed two weeks after his election success on December 3, 2006, fearing the development of revolutionary consciousness among the workers. To one opponent's statement that he was in favor of a 'democratic socialism', Chávez replied that the problem was that 'I am a socialist and he is a social-democrat', and he added, 'I am in favour of revolutionary socialism' (cited in ibid.). Actual membership of, not merely voters *for*, the PSUV is estimated at 5.8 million (Venezuelanalysis.com, 2007).

In talking about the need for a revolutionary leadership Chávez also quoted from Lenin on the need for a revolutionary party in order 'to articulate millions of wills into one single will', which 'is indispensable to carry out a revolution, otherwise it is lost, like the rivers that overflow, like the Yaracuy that when it reaches the Caribbean loses its riverbed and becomes a swamp' (cited in ibid.). Chávez argued that PSUV must be the most democratic party Venezuela has ever seen, built from the bottom up, inviting all the currents of the Venezuelan left to join.

He also insisted that it must not be dominated by electoral concerns, nor by the existing leaders of the existing coalition parties. He criticized the way the Bolshevik Party in Russia came to suffocate rather than stimulate a battle of ideas for socialism, noting how the marvelous slogan of 'all power to the soviets' degenerated into a sad reality of 'all power to the party'. For Chávez, this points towards precisely the kind of mass, democratic, revolutionary, political organization that is needed (Piper, 2007b).

At the time of writing (Summer 2008), in front of an audience composed mainly of PSUV leaders, Chávez has just made a further reference to Trotsky, in commending a book by UK Marxist writer Alan Woods (Woods, 1999) (Corriente Marxista Revolucionaria [Venezuela] 2008)

Here is the quote:

> [The class struggle] needs a correct program, a firm party, a trustworthy and courageous leadership—not heroes of the drawing room and of parliamentary phrases, but revolutionists, ready to go to the very end. (cited in ibid.)

Chávez said that the PSUV leaders should adapt themselves to Trotsky's phrase (ibid.).

Capitalists and their political supporters are intent on spreading disinformation about the Chávez Government.[17] In particular, there are numerous attempts to label the Government nondemocratic or 'dictatorial'. In actual fact, *Latinobarómetro*'s, 2008 poll, surveying the development of democracy, economies and societies in Latin America, revealed that Venezuela is ranked second in terms of how satisfied citizens are with democracy, and as the Latin American country where the highest percentage of its citizens describe their economic situation as positive (Venezuela Information Centre, January 21, 2008, venezuela@btconnect.com).

As Jorge Martin (2007) concludes:

> [t]he political thinking of Chavez is in tune and reflects the conclusions drawn by tens of thousands of revolutionary activists in Venezuela, in the factories, in the neighbourhoods, in the countryside. They are growing increasingly impatient and want the revolution to be victorious once and for all.

Antiracism in Practice

In the light of CRT concerns about 'white Marxism' (see, for example the critique of Mills' and Gillborn's views on this in chapter 5 of this volume), it is worth pointing out that Chávez was the first Venezuelan President ever to claim and honor his indigenous and African ancestry.[18] It is also important to emphasize the antiracist developments currently occurring in Venezuela. Chávez articulated this when he stated:

> We've raised the flag of socialism, the flag of anti-imperialism, the flag of the black, the white and the Indian...I love Africa. I've said to the Venezuelans that until we recognise ourselves in Africa, we will not find our way...We have started a hard battle to bring equality to the African descendents, the whites and the indigenous people. In our constitution it shows that we're a multicultural, multiracial nation. (Chávez, 2008, cited in Campbell, 2008, p. 58)

Like the rest of Latin America, Venezuela's history is scarred by colonialism's and imperialism's racist legacies. Only now, with the gains being made by the Chávez Government and the growing mass revolutionary movement, is Venezuela beginning to grapple in earnest with how to confront this racist legacy.

The rights of Venezuela's indigenous people were first entrenched in the 1999 Bolivarian constitution (as noted earlier in this chapter, Chávez came to power in 1998), which was ratified by 71% of voters. For the first time, indigenous land rights were identified as being collective, inalienable, and nontransferable, recognizing the

> rights of the indigenous peoples over the land they traditionally and ancestrally occupied. They must demarcate that land and guarantee the right to its collective ownership. (cited in Harris, 2007)

As Harris points out:

> Article 9 stipulates that while Spanish is Venezuela's primary language, 'indigenous languages are also for official use for indigenous peoples and must be respected throughout the Republic's territory for being part of the nation's and humanity's patrimonial culture'. The 1999 constitution also affirms that 'exploitation by the state of natural resources will be subject to prior consultation with the native communities', that 'indigenous peoples have the right to an education system of an intercultural and bilingual nature', that indigenous people have the right to control ancestral knowledge over 'native genetic resources' and biodiversity, and that three indigenous representatives are ensured seats in the country's National Assembly. (these were elected by delegates of the National Council of Venezuelan Indians in July 1999)

Since 1999, the confidence of the indigenous rights movement has exploded. The multitude of social problems that persist as a hangover of previous, capitalist policies has led to a culture of Chávista activists who support the revolution and lobby the Chávez government to demand attention to their particular issues (Harris, 2007).

One organization at the forefront of the antiracist movement is the Afro-Venezuelan Network, headed by Jesus 'Chucho' Garcia, which is lobbying for recognition of Afro-Venezuelans in the next round of amendments to the Bolivarian constitution. This Network successfully campaigned for the creation of a presidential commission against racism in 2005, the inclusion of Afro-Venezuelan history in the school curriculum, the establishment of a number of cocoa-processing plants and farming cooperatives run by black Venezuelans and for Afro-Venezuelan Day on May 10 of each year (Harris, 2007).

As Harris (2007) explains, the ambitious land and agrarian reforms embedded in the 1999 constitution have been especially beneficial to indigenous and Afro-Venezuelan communities. The constitution declares that idle, uncultivated private land over a certain size can be transformed into productive units of land for common social benefit. 'By prioritizing socially productive land use over monopolistic private land ownership and redistributing idle land to the landless, Chávez has promoted independence, food sovereignty and local agricultural development' (Harris, 2007).

Such developments are not confined to Venezuela. Chávez has also been building alliances with other marginalized communities in the Americas, including providing food, water and medical care to 45,000 Hurricane

Katrina victims in areas surrounding New Orleans, and supplying discounted heating and diesel oil to schools, nursing homes and hospitals in poor communities in the United States (Harris, 2007).

Harris (2007) concludes:

> in Venezuela the space for frank discussion about how to move forward in the context of a mass movement has been opened up by the ongoing revolutionary process, and genuine gains have been made by indigenous and Afro-Venezuelan movements to eliminate the systemic nature of racism from Venezuelan life.

Whatever the eventual outcome, the current climate and developments in the Bolivarian Republic of Venezuela should provide hope for the future for Marxists and Critical Race Theorists alike. However, these developments cannot be clearly understood or analyzed by CRT, unless articulated with Marxist analysis.

In this chapter I began by arguing that capitalism has an inbuilt tendency to generate conflict. I went on to give a few examples of potentially revolutionary situations, noting the success of capitalist ideology in keeping Marxism off the agenda. I then considered a number of common objections to Marxism, and gave my own Marxist response. In the final part of the paper, I looked at developments in the Bolivarian Republic of Venezuela as examples of twenty-first social democracy in practice and the possibility of socialism in embryo. Contra the CRT assertion that Marxism is 'an exercise of White power' (as Gillborn interpreted Allen in chapter 5 of this volume), I argued that, in Venezuela, serious attempts are being made to grapple with that country's racist colonial legacy and to move forward in a socialist direction. I hope I have now fully laid the groundwork for the final chapter, in which I will direct my attention to some implications for classroom practice of CRT and Marxism respectively.

Chapter 8

CRT and Marxism: Some Suggestions for Classroom Practice

Throughout this book, I have dealt with a large number of issues pertaining to CRT, and while acknowledging some of CRT's strengths, I have also highlighted the essential differences between CRT and Marxism. In this final chapter, I begin by noting some suggestions for classroom practice from Critical Race Theorists with which Marxist educators would be in general agreement, but add some caveats. I then go on to critique some classroom pedagogies, based on 'the abolition of whiteness', and advocated by a leading UK Critical Race Theorist. Next I make some suggestions for promoting equality in the school classroom, based on Marxism. Specifically, I address the issues of *antiracist multicultural education; the reintroduction of the teaching of imperialisms;* and what I have called *the last taboo: the teaching of democratic socialism,* including *ecosocialism.* I conclude the chapter by making suggestions as to where such discussions might take place. While my examples are drawn from the UK classroom, I believe they can be applied to other educational settings.

Some Areas of Agreement

While Marxists would disagree with certain assertions, for example racism is permanent and central, I have broad agreement with the suggestions for classroom practice of Daniel and Tara Yosso (2005, pp. 70–72). Their brief is higher education, but it is my view that these suggestions apply equally to elementary/primary/junior and secondary/high schools. Solórzano and Yosso (p. 70) argue that 'race' and racism should be discussed in the classroom, that racism intersects with other forms of oppression, and that racism is not a black/white binary (p. 70).

They go on to stress that CRT in the classroom must challenge the dominant ideology of 'objectivity, meritocracy, color-blindness, race neutrality, and equal opportunity' (ibid.). Critical Race Theorists, they argue, are committed to social justice and liberation with respect to 'race', gender and class (p. 71). To this, I would wish to add sexual orientation and disability and other forms of oppression.

I am, of course, in full agreement with their wish to eliminate poverty and empower underrepresented groups (ibid.), and with their acknowledgment that educational institutions 'operate in contradictory ways, with their potential to oppress and marginalize coexisting with their potential to emancipate and empower' (ibid.). To their argument that the experiential knowledge of people of color is 'legitimate, appropriate, and critical to understanding, analyzing, and teaching about racial subordination' (ibid.), I would want to incorporate, as argued in chapter 3 of this volume, Leonardo's (2004) call to integrate this focus on subjectivity with Marxist objectivism. I would further wish to insist that the voice and experiential knowledge of those racialized on non-color-coded grounds (see chapter 2 of this volume) is also legitimate, appropriate and critical, as is indeed the experience and voice of the exploited and oppressed non-racialized white working class pupils/students (and I use 'working class' in its sociological sense here).

Finally, Solórzano and Yosso (2007, p. 70) claim that CRT 'challenges ahistoricism' and is transdisciplinary With respect to 'ahistoricism', I have suggested in chapter 5 of this volume that not all CRT analyses incorporate this challenge. As far as a transdiciplinary knowledge base is concerned, I would fully endorse the need to utilize various curriculum subject areas to understand and undermine racism and other inequalities. Indeed, I have attempted just this in Cole (2009b) (see chapter 4, note 14 of this volume).

I would also agree with Marvin Lynn (2007, p. 131), in his discussion of 'Critical Race Theory and its Links to Education' that there is a need 'to look analytically at the failure of the educational system . . . to properly educate the majority of culturally and racially subordinated students', and to examine 'racial sorting' where schools put pupils/students of color in lower tracks (streams), are over-represented in 'special' schools and pushed out of certain schools (ibid.). But again, I would want to stress that certain pupils/students who are racialized in non-color-coded ways may be on the receiving end of similar processes, and that non-racialized white working class (again in the sociological sense) pupils/students also get 'sorted' in ways that are to their detriment (Lynn also refers to this, but only in passing (ibid.)). I would further agree with Lawn (ibid.) when he refers to 'the systematic annihilation of Black and Brown students' through the whole education system (ibid.), although I would use the phrase 'structural and systematic' and would include the other constituencies I refer to immediately above.

Lynn concludes his discussion of 'CRT and Education' by citing Solórzano and Yosso's (2000, p. 42) claim that CRT in education 'seeks to identify, analyze, and transform those structural, cultural, and interpersonal aspects of education that maintain the subordination of Students of Color', and, in addition, that it 'asks such questions as: what roles do schools themselves, school processes, and school structures play in helping to maintain racial, ethnic and gender subordination' (Solórzano and Yosso, 2000, p. 40, cited in Lynn, 2007, p. 131). Finally Lynn suggests, also following Solórzano and

Yosso, 2000 that 'CRT can be utilized as a point from which to begin the dialogue about the possibilities for schools to engage in the transformation of society' (Lynn, 2007, p. 131). The transformation of society is of course the guiding principle of Marxism, but whereas, for Marxists, transformation entails a social revolution and transition to socialism, Critical Race Theorists in general, like poststructuralists and postmodernists in toto (see Cole, 2008d, Chapter 5), lack a clear vision of a transformed society or, indeed, a transformed world. Marx was suspicious of philosophers (and I am sure would be equally suspicious of Critical Race Theorists) who had 'interpreted the world in many ways', whereas for him, the point was 'to change it' ([1845] [1976], p. 123). I return to a discussion of this limitation of CRT in the Conclusion to this volume. For a discussion of Marxist ideas about socialist transformation, and a consideration of twenty-first century socialism, see chapter 7 of this volume.

Lynn (2007, pp. 137) concludes his analysis in typical CRT fashion by stressing the existence of white supremacy 'in the United States and in the world'. With respect to the notion of 'white supremacy' per se as a useful descriptor of everyday racism I have critiqued this in chapter 2 and elsewhere in this volume. As far as claims for its ubiquitous presence are concerned, I have also argued that 'white supremacy' is a most valid *historical* descriptor for certain societies but not an informative current one, and also that world racism takes a number of forms which are not related to skin color.

Preston's Classroom Pedagogies: A CRT Strategy[1]

In chapter 2 of this volume, I discussed 'the abolition of whiteness', 'white supremacy' (and attendant 'white privilege'). While Vaught and Castagno (2008, p. 102) argue that understanding white privilege must be an element of 'teacher training', John Preston makes the case for the 'abolition of whiteness' to part of classroom pedagogy. 'Abolition of whiteness' needs to be anchored theoretically within the broader 'Whiteness Studies' movement. Preston (2007, p. 6), following Byrne (2006) and Hill (1997), describes two phases of work on Whiteness Studies The first is the 'pre-political' phase, where whiteness is seen to be just another form of identity, from which certain social advantages arise. Writers in the first phase were mainly scholars of color. Preston lists Franz Fanon, Du Bois, and Sojourner Truth among others (p. 7).

In the second phase, discussed in chapter 2 of this volume, 'whiteness becomes a political category in that a structural apparatus of oppression is posited' (ibid.), entailing white privilege within an overarching system of 'white supremacy' (ibid.). In this second phase, not only is whiteness interrogated as an identity, but the consequences of 'acting white' are made visible' (ibid.). In this second phase, white supremacy is considered critically only in the sense of white identity and white privilege (ibid.). Preston cites the work of Peggy McIntosh (McIntosh, 1988) (discussed earlier in this volume) as being part of this phase, as well as Cheryl Harris (1998) (also discussed earlier in this volume). In this second phase, Preston concludes, Whiteness

Studies became more 'critical' in that whiteness is considered not just an 'identity', but an identity where certain practices lead to 'white privilege'.

Preston then identifies a third phase (p. 8). In this phase, there is both a critique questioning the basis and direction of the second phase, and 'wish to radicalize it towards the abolition of whiteness or white supremacy' (pp. 8–9). It is this phase which is most associated with CRT, and this phase on which Preston's classroom pedagogies are based.

With respect to Preston's choice of the word, 'pedagogy', Terry Wrigley and Peter Hick (forthcoming, 2009) have argued that the word *pedagogy* is relatively new in English-speaking countries, and that it is often used with limited understanding. In most European languages and education systems, they argue the concept means more than just teaching methods. It requires 'an articulation of educational aims and processes in social, ethical and affective as well as cognitive terms, and involves reflection about the changing nature of society or the value of human existence' (Wrigley and Hick, forthcoming, 2009).

It is the European sense that Preston (2007) seems to be adopting since, he attempts to provide a way in which such 'critical theory' might be introduced into schools. Preston (2007, p. 198) concludes his book by advocating neo-abolitionist (the word is prefixed with 'neo' to differentiate the abolition of whiteness from the earlier abolition of slavery) pedagogies ('abolition of whiteness' teaching) in the classroom. Given that the undoubtedly good intentions of the 'abolition of whiteness' arguments are regularly misunderstood by academics (Preston, 2007, passim), its introduction in the school curriculum is a most worrying and counter-productive suggestion. That whiteness (not 'white people') should be abolished is advocated by Preston for the following reasons, based on the work of Noel Ignatiev and John Garvey:

1. 'whiteness is a false form of identity and...there is no such thing as white culture';
2. 'whiteness, in terms of a structural system of white supremacy, is oppressive...[and] whiteness is *only* false and oppressive and...there is no possibility of 'redemption' or reformation of whiteness';
3. 'whiteness divides humanity against itself and therefore is not in the genuine interests even of white people';
4. 'class, gender and sexuality are important in understanding oppression but race is central to understanding why other forms of political activity are not possible, particularly in the US' (Preston, 2007, p. 10)

I am not sure what Preston means by proposition 4, but will consider each of these other propositions in turn.[2]

Whiteness Is a False Form of Identity and...There Is No Such Thing as White Culture

While I agree that there is no such thing as 'white culture' per se, there are white *cultures*. It is particularly important, given the scenario of continuing

UK white working class racism (exacerbated, as I have argued throughout this volume by sections of the tabloid press), that educators do not deny the existence of white working class cultures. Indeed, as I have argued elsewhere with respect to such cultures (Cole, 2007c, 2008c), educational institutions should be centrally involved in helping to identify and develop strategies to promote good inclusive practice for *all* pupils/students, including the white working class, *non-racialized* as well as racialized (see below). Sections of the white working class in England have voted for the fascist British National Party (BNP) at recent elections *precisely* because they feel that they are treated with less equality than others. If we were to *teach* white working class young people that they have no culture, or indeed if we were to treat them as if they had no culture, that would be racist, would alienate white working class children even more, and would not be conducive to effective socialist practice. The notion of such a lack of culture, which would surely lead to identity crises (a point that fellow 'white abolitionist' Ricky Lee Allen (2007, p. 65) seems to revel in when he states that critical educators need to create an environment which creates this) would also rightly be massively contested, including by most of the Left in the United Kingdom.

Whiteness Is a Structural System of Oppression and There Is No Possibility of Redemption or Reformation of Whiteness

I would argue that it is *capitalism* not white supremacy that is a structural system of oppression. With capitalism's overthrow, there is every possibility that the color of one's skin will be irrelevant and racism (which, as I have argued, is not necessarily based on skin color) abolished. While it may be the *intention* of Critical Race Theorists to make skin color irrelevant, it is my view that encouraging young people in schools to think on these lines is also not conducive to effective socialist practice. In chapter 5 of this volume, I referred to current developments in Venezuela (see also Cole, 2008g, 2008h) which point to a revolutionary process where whiteness is neither redeemed, nor reformed nor abolished but, *in the context of major ameliorative projects,* seen as a constituent form of identity in an antiracist struggle for twenty-first century socialism. For Chavez, as I noted in that chapter, 'the flag of the black, the white and the Indian' has been raised (cited in Campbell, 2008, p. 58).

Whiteness Divides Humanity against Itself and Therefore Is Not in the Genuine Interests Even of White People

A belief that a division of 'whiteness' divides humanity is not surprising, given Preston's claims that whiteness is an objective power structure. As noted above, for Chakrabarty and Preston (2006, p. 1) white supremacy, along with capitalism, is an objective inhuman system of exploitation and oppression. From a Marxist perspective, it is capitalism that is the objective system that divides humanity against itself, and is against the interests of *all* workers. The Labor Theory of Value (LTV) explains most concisely why capitalism is objectively a system of exploitation, whether the exploited realize it or not, or indeed, whether they believe it to be an issue of importance for them or not. The LTV also provides a *solution* to this exploitation. It thus provides *dialectical* praxis—the authentic union of theory and practice. According to the LTV, the interests of capitalists and workers are diametrically opposed, since a benefit to the former (profits) is a cost to the latter (Hickey, 2002, p. 168) (An explanation of the LTV is provided in the appendix to this chapter).

Some Suggestions for Classroom Practice, based on Marxism

In contemporary societies, we are in many ways being globally *mis*educated. The Bush and Blair administrations' propaganda war about "weapons of mass destruction", aimed at masking new imperialist designs and capital's global quest for imperial hegemony and oil, was a key example.

Conditioning the discourse is only half the story. 'Education' has become a key component in the profit-making process itself. Tied to the needs of global, corporate capital, 'education' world-wide has been reduced to the creation of a flexible workforce, the openly acknowledged, indeed lauded (by both capitalists and politicians) requirement of today's global markets. Corporate global capital is in schools, both in the sense of determining the curriculum and exercising burgeoning control of schools as businesses. In the United Kingdom, capitalist infiltration of education has been given a major boost by the new Gordon Brown government.

An alternative vision of education is provided by Peter McLaren. Education should, McLaren argues, following Paulo Freire, put 'social and political analysis of everyday life at the centre of the curriculum' (McLaren, 2003, p. xxix). Marxists are clear as to their role is in the debate over the future of our planet, and of the way education must be central in working towards a better future. As Madan Sarup (1988, pp. 147–148) argues:

> A characteristic of human beings is that they make a distinction between the 'real' and the 'ideal'…Human beings have a sense of what is possible in the future and they have the hope that tomorrow will be better than today. Marxists not only have this hope, this orientation towards the future, but they try to understand the world, to develop a critical consciousness of it, and try to develop strategies for changing it. Of course, they realize that progress is uneven, not unilinear; because of the nature of contradiction there are inevitably negative aspects, sad reversals and painful losses. Marxists struggle for a better future for all, but they know that this does not mean that progress is guaranteed or that the processes of the dialectic will lead to the Perfect… education is closely connected with the notion of a change of consciousness; gaining a wider, deeper understanding of the world represents a change for the better. And this, in turn, implies some belief in a worthwhile future. Without this presupposition the education of people would be pointless.

Racism should be a key component in education to change consciousness. I would argue that, in order for racism to be understood, and, in order for strategies to be developed to undermine it, there is a need first to promote a thorough antiracist multicultural education, which is cognizant of the manifestations of xeno-racism and xeno-racialization; second to reintroduce the topic of imperialism in schools and other educational establishments; third, Marxists should make the case for the teaching of democratic socialism (which incorporates ecosocialism). I will deal with each in turn.

Antiracist Multicultural Education

I have discussed Islamophobia throughout this volume. It is important to also emphasize ongoing antiblack and anti-Asian racism. Indeed, I noted in chapter 5, when referring to the research of Gillborn, how black students and their peers of Pakistani and Bangladeshi heritage continue to be significantly less likely to achieve the key benchmarks when compared to white peers of the same gender. Moreover, the most recent research (Strand, 2008, cited in *The Guardian*, September 5, 2008, p. 3) shows that black Caribbean children are being subjected to institutional racism that can dramatically undermine their chances of academic success. I also referred, in chapter 2, to the underachievement of Gypsy, Roma and Traveller (GRT) students. I have stressed the need to combat all forms of non-color-coded racism.

Having consistently argued in the past (e.g., Cole, 1986a, 1986b, 1989, 1998b, 1998c, 1992b, c) *against* multicultural education, and *for* antiracist education, I now believe that technological advances, in allowing people to speak for themselves—for example via websites, blogs and email—provide a window of opportunity for multiculturalism; and thus for antiracists to modify their position to include multicultural education. The way forward, I believe, is to promote both antiracism and antiracist multiculturalism: to create opportunities for hearing authentic voices, but, as argued in chapter 2, in conjunction with objective analysis. I would thus now advocate an *antiracist* multicultural education. This should avoid simplistic versions of 'Racist Awareness Training' (RAT), practiced in the past in the United Kingdom, whereby *all* white people were considered to be infected with a racist virus which could be cured by the right therapy (for a critique of RAT, see Sivanandan, 1985).[3]

Using the web creatively, antiracist multicultural education should be about the importance of antiracism as an underlying principle, about the need to constantly address institutional racism and 'race' equality, and about the promotion of respect and nonexploitative difference in a multicultural world. As we saw in chapter 4 of this volume, during the 1970s and 1980s, and into the 1990s in Britain, there was a protracted debate between those, broadly liberals, who wished to promote multicultural education (celebrating the diversity of cultures which make up British society), and those, mainly politically Radical Left, who advocated antiracist education (viewing the institutional racism of British society as the fundamental problem).

In terms of actual hegemonic practice in schools, most schools were and have remained monocultural (promoting 'British culture and values', whatever that may mean), some have practiced multicultural education, and, only a few have actually put antiracist education into practice.

As we also saw in chapter 4 of this volume, the antiracist critique of mono-cultural education is that in denying the existence of, or marginalizing the cultures of various minority ethnic communities, it was and is profoundly racist. The antiracist critique of multicultural education is that it was and is both patronizing and superficial, and often offensive. In chapter 4, I also tried to demonstrate the Marxist underpinnings of my favored form of anti-racist education in my example of the Australian bi-centennial of 1988.

The Stephen Lawrence Inquiry Report (Macpherson, 1999), as I argued in chapter 4, was a milestone for antiracists, in being the first acknowl-edgement by the British State of the existence of widespread institutional racism. Antiracist multicultural education must have, as well as antiblack racism and Islamophobia, non-colour-coded racism as integral to its pro-ject. It is highly likely that xeno-racism has filtered down into schools in the United Kingdom, and is impacting on day-to-day pupil interaction. How this will affect inter- and intra-ethnic relations is largely unforeseen. What is clear is that anti-xeno-racist multicultural education must feature largely in UK schools' priorities. There is also some evidence of racism from stu-dents from Newly Joined Countries of the European Union (NJCEU) pupils being directed at Asian and black pupils in the United Kingdom (Glenn and Barnett, 2007), and this clearly needs addressing too. A good example of antiracist multicultural education (which incorporates anti-xeno-racism) in practice is Filton High School in Bristol in the United Kingdom (see Vernell and Carter, 2008).

The Reintroduction of the Teaching of Imperialisms

Anti-imperialism is one of Chávez's main platforms. As he remarked in 2003:

> In Venezuela, we are developing a model of struggle against neoliberalism and imperialism. For this reason, we find we have millions of friends in this world, although we also have many enemies (cited in Contreras Baspineiro, 2003).[4]

I have dealt with the teaching of imperialism in schools at length elsewhere (e.g., Cole, 2004d). Here I will make a few general points. Reintroducing the teaching of imperialism in schools I believe, would be far more effective than CRT in increasing *an understanding* of the nature of racism, and cru-cially linking racism to capitalist modes of production.

Students will need skills to evaluate the New Imperialism and 'the perma-nent war' being waged by the United States with the acquiescence of Britain. Boulangé (2004) argues that it is essential at this time, with the Bush and Blair 'war on terror', and Islamophobia worldwide reaching new heights, for teachers to show solidarity with Muslims, for 'this will strengthen the unity of all workers, whatever their religion' (p. 24), and this will have a powerful impact on the struggle against racism in all spheres of society, and education in particular. In turn, this will strengthen the confidence of workers and students to fight on other issues.

According to neo-Conservative, Niall Ferguson (2003):

> Empire is as 'cutting edge' as you could wish...[It] has got everything: economic history, social history, cultural history, political history, military history and international history—not to mention contemporary politics (just turn on the latest news from Kabul). Yet it knits all these things together with...a 'metanarrative'

For Marxists, an understanding of the metanarrative of imperialism, past and present, does much more than this. Indeed, it encompasses but goes beyond the centrality of 'racial' liberation in CRT theory. It takes us to the crux of the trajectory of capitalism from its inception right up to the twenty-first century; and this is why Marxists should endorse the teaching of imperialism old and new. Of course, the role of education in general, and teaching about imperialism in schools in particular, has its limitations and young people are deeply affected by other influences and socialized by the media, parents/carers, and by peer culture (hence also the need for media awareness). Unlike Marxism, CRT does not explain why Islamophobia, the 'war on terror' and other forms of racism are necessary to keep the populace on task for 'permanent war' and the accumulation of global profits.

The Last Taboo: The Teaching of Democratic Socialism in Schools

Chávez devoted a May 15, 2005 call-in television program to education. In direct contrast to the U.S. and the UK view that we should teach the entrepreneurial culture in schools, for Chávez there is a new educational model: unity and solidarity must replace competition and individualism in schools. As he put it, '[w]e are all a team, going along eliminating little by little the values or the anti-values that capitalism has planted in us from childhood' (Chávez, cited in Whitney Jr., 2005).

As noted above, capitalist inroads into UK schools, colleges, and universities have been given a boost by the new Gordon Brown government. Indeed, New Labour's promotion of capitalist values in the education system has intensified to such an extent that the stance of Thatcher and the Radical Right in the 1980s seems tame by comparison. If we are to have 'corporate values' in the education system, if we are to have 'business-facing' universities, then it seems quite reasonable to press for the introduction in education of alternatives to neoliberal global capitalism and imperialism, such as (world) democratic socialism. No space in the education systems of the United States and the UK is provided for a discussion of alternatives. Marxists should agitate for the (totally democratic) suggestion that such discussions should take place in schools, colleges and universities.

Students would benefit from engaging in an analysis of the mechanics of capitalist production and exchange. Marxism would be an obvious starting point. Such an analysis should have as central a discussion of the LTV (see the appendix to this chapter) since this most clearly explains exactly *why* Marxists believe that capitalism is *objectively* a system of exploitation (the teaching of the LTV was, in fact, compulsory in secondary schools in the former Yugoslavia). Students could consider the concept of globalization. Is it a new phenomenon, or is it as old as capitalism itself? Is it inevitable, as claimed by many (see chapter 6 of this volume). To what extent is the concept of globalization ideological? Does it hide more than it reveals?

As outlined in chapter 6 of this volume, neoliberal capitalism, in being primarily about expanding opportunities for large multinational companies, has undermined the power of nation states and exacerbated the negative effects of globalization on such services as healthcare, education, water and transport. The effects on both the 'developed' and 'developing world' should be discussed openly and freely in the classroom.

Ecosocialism

A consideration of democratic socialism should incorporate a thorough consideration of ecosocialism. McLaren and Houston (2005, p. 167) have argued that 'escalating environmental problems at all geographical scales from local to global have become a pressing reality that critical educators can no longer afford to ignore'. They go on to cite 'the complicity between global profiteering, resource colonization, and the wholesale ecological devastation that has become a matter of everyday life for most species on the planet' (ibid.) (see chapter 6 for a discussion). Noting the wealth of ecosocialist scholarship that has emerged in recent years (e.g., Williams, 1980; Benton, 1996; Foster, 2000, 2002; see also Feldman and Lotz, 2004), McLaren and Houston, following Kahn (2003), state the need for 'a critical dialogue between social and eco-justice' (p. 168). They call for a dialectics of ecological and environmental justice to reveal the malign interaction between capitalism, imperialism and ecology that has created widespread environmental degradation which has dramatically accelerated with the onset of neoliberalism (p. 172; see chapter 6 of this volume for a discussion of these interconnections; see also Cole, 2008d, Chapter 7). McLaren and Houston (p. 174) then propose an educational framework, of which the pivot is class exploitation, but which also, following Gruenwald, 2003, interrogates the intersection between 'urbanization, racism, classism, sexism, environmentalism, global economics, and other political themes'. The classroom is a good arena to discuss issues, ranging from what is happening in the immediate vicinity of the school, to issues at the national policy level, through to global issues, including the ecosocial issues connected to the global survival of the planet. Students could begin by discussing the issues discussed chapter 6 of this volume: the destruction of resources; unhealthy food; genetic modification; and climate change. They could then interrogate the causes, and assess the likely chances of changes under neoliberal capitalism and the 'New Imperialism' (see chapter 6 of this volume)—for example, what can be done *now* to address these pressing issues, and how a world socialist system might do things differently.

Where Might These Discussions Take Place?

Linking Up with the Community

It is important for students to link up with local communities. Students need to be aware of how racism affects local community groups and of local struggles for better jobs, working conditions, health services, day care facilities, housing and so on.

It is important to *interact* with local racialized communities, in the pursuit of reciprocal knowledge. It is absolutely crucial, in my view, for organizations representing racialized groups to link up with each other *and* with non-racialized white working class communities, and to develop strategies for reciprocal mutual support and well being and to undermine the growth of organizations like the British National Party (BNP) whose tactics are to divide rather than unite communities. Residents' associations and youth organizations might also be good inroads, in defending and enhancing the rights of *all* the members of the local community.

Linking up with communities facilitates an understanding of racism, Islamophobia and xeno-racism, locally, nationally and internationally. Understanding processes of racialization and xeno-racialization will inform the ways by which people become demonized because of their (perceived) identities.

It needs to be stressed that working with local communities requires considerable skills from Marxist educators or the purpose of such links might well backfire. Thus it will be important initially to make links with Left activists and other equality advocates in the various communities in order to decide strategies. A joint meeting with these various activists and advocates might be a good starting point.

The UK National Curriculum

I will conclude with some *specific* suggestions as to where in the current UK National Curriculum, these issues might be most obviously addressed. While my focus is the UK, the issues discussed are universally applicable.

The Global Gateway

The UK government's international strategy and the DfES Global Gateway (www.globalgateway.org) provides the opportunity to register schools and to link up with schools worldwide. Referring to its potentials, Olga Stanojlovic, director of communications for the British Council's education and training group, which developed and maintains the 'Global Gateway', suggests that British schools might link up with schools in Africa 'to learn more about sustainable development of different ideas of social justice and citizenship around the world' (cited in the *Education Guardian*, 14 November, 2006: 1). This is a most worthwhile suggestion. However, if we are to be fully honest with the young people in our schools, we must link them up in addition with schools in Venezuela, Cuba, Bolivia and elsewhere for a *comprehensive* analysis of ideas about social justice. We must initiate discussions about (world) socialist alternatives to imperialism and to neoliberal global capitalism. The *British Venezuela Solidarity Campaign* website (http://www.venezuelasolidarity.org.uk/) is a useful resource, as is *Hands Off Venezuela* (http://www.handsoffvenezuela.org/home/).

The Revised Citizenship Curriculum at KS3/4

According to the QAA:

> The revised programme of study [at KS3] recognises the importance of engaging pupils in thinking through, and responding to, real dilemmas, issues and problems facing individuals and communities. It encourages pupils to develop new ways of thinking and reflecting on a wide range of issues, ideas and concepts, including democracy and justice (QAA, 2007a).

The QAA goes on to define 'democracy and justice' as involving 'understanding the underpinning values and processes upon which *our* society is based, including freedom, fairness and equality before the law' (my emphasis). However, it then states that, '[s]uch an understanding requires independent enquirers and effective participators who are able to make judgements about the extent to which decisions, action and opinions reflect these important values' (QAA, 2007b). Here there is space for comparing with British capitalist 'democracy' with socialist democracy.

At KS4, Citizenship Education, according to the QAA, should help 'young people to develop their critical skills and to consider a *wide range* of political, social, ethical and moral problems and explore opinions other than their own' (my emphasis) (QAA, 2007c). This also more than lends itself to a consideration of the socialist alternative.

Every Child Matters

The five aims of the UK government's *Every Child Matters* (DfES, 2004a, b, 2005) are for children to:

• Be healthy
• Stay safe

- Enjoy and achieve
- Make a positive contribution
- Achieve economic well-being.

While these are certainly laudable aims, they might be seen as rhetorical in the light of the actual material and psychological situation of Britain's young people after the decade of the Blair government. In a UNICEF report of twenty-one industrialized countries issued on St. Valentine's Day (UNICEF, 2007), the UK came joint-last with the United States for 'child well-being' (see Cole, 2008h for a discussion). That this is the case and why it is the case should be part of schools' implementation of *Every Child Matters*. Introducing dialogue on this in schools needs to make links to the capitalist economy and capitalist priorities.

In addition, while *Every Child Matters*' aims are clearly concerned with the welfare of children and young people in the UK, there is scope here for adding an international dimension. For example, under 'being healthy', children and young people could learn about global health issues, and examine what eating healthily means to children in the 'developing world' (Smith, cited in Jewell, 2006: 4). For 'staying safe' they could learn that safety for children is enshrined in human rights, but that safety depends on where children and young people live, and the political and economic context in which they find themselves (Smith, cited in ibid.). For 'enjoy and achieve', children and young people can realize that achievement can transcend personal success and can contribute to making a difference in society locally, nationally, and globally, and that this in itself can bring joy. For 'achieve economic well-being', they can learn about global capitalism, and 'how we are rich in the West because other areas of the world are poor' (cited in ibid.). With respect to this point, while it is of course demonstrably true, ownership of wealth in the west is relative, as witnessed by the UNICEF (2007) report, and many people are very poor in the UK too. Finally, 'making a positive contribution' could empower young people by their getting involved in communities locally, nationally, and internationally, with a view to campaigning for social justice in all its forms, in the context of the protection and care of the local, national, and international environment. Such a model stresses young people as active participants rather than merely citizens of the future. The National curriculum and other documents tend to emphasize 'preparation for adult life', which is more patronizing than empowering.

Internationalizing *Every Child Matters* should crucially include an examination of what such a concept has meant for nearly half a century in Cuba, and what it means in the context of the Bolivarian Revolution in Venezuela, in the context of developments in Bolivia and elsewhere. Internet makes this eminently possible.

In this chapter I begin by critiquing some classroom pedagogies which are informed by CRT. I then went on to make some suggestions, based on Marxism, for promoting equality in the school classroom. I concluded the chapter by making suggestions as to where such discussions might take place.

In the Conclusion to the book, I will try to draw some of the arguments of this book together.

Appendix

Marx argued that workers' labor is embodied in goods that they produce. The finished products are appropriated (taken away) by the capitalists and eventually sold at a profit. However, the worker is paid only a fraction of the value s/he creates in labor; the wage does not represent the *total* value s/he creates. We *appear* to be paid for every single second we work. However, underneath this appearance, this fetishism, the working day (like under serfdom) is split in two: into socially necessary labor (and the wage represents this) and surplus labor, labor that is not reflected in the wage. Greatly oversimplifying matters, let us assume that a capitalist employs a worker to make a table. Let us say that the value of the basic materials is £100, and that after these basic materials have had labor embodied in them (i.e., have become a table) that table has a value of £500. Let us further assume that in the time it takes to make the table £20 of overheads is used up. What happens to the £400 surplus value that the worker has created? The worker is paid, say £100 and the remaining £300 are appropriated, is taken away, by the capitalist. After overheads are paid, the capitalist still has £280 *surplus* that he or she can reinvest to create more surplus. To continue the example, with this £280 surplus the capitalist can buy £200 worth of basic materials, and employ two workers, and after these basic materials have had labor embodied in them (e.g., have become two tables) those tables have a value of £1000. Assuming overheads increase to £30, and two workers are each paid £100, the capitalist is now left with £770 surplus which can be thrown back into production to create yet more surplus value, and so on and so on. If the capitalist continues to employ workers, say seven, the surplus would be over £6000. It is thus easy to see how surplus value multiplies and how capitalists' surplus (which is converted into profit) is, in truth, nothing more than accumulated surplus value, really the 'property' of the worker but appropriated from that worker. (Marx argues that the origins of the capital held by capitalists lies in the forcible seizure of feudal and clan property, the theft of common lands and state lands, and the forced acquisition of Church property at nominal price. In other words, capitalism has its origins in theft and continues on the same basis (see Marx, 1965 [1887], pp. 717–733).

While the value of the raw materials and of the depreciating machinery is simply passed on to the commodity in production, labor power is a peculiar, indeed unique commodity, in that it creates new value. 'The magical quality of labour-power's...value for ...capital is therefore critical' (Rikowski, 2001, p. 15). '[L]abour-power creates more value (profit) in its consumption than it possesses itself, and than it costs' (Marx, 1966 [1894], p. 351). Unlike, for example, the value of a given commodity, which can only be realised in the market as itself, labor creates a new value, a value greater than itself, a value that previously did not exist. It is for this reason that labor power is so important for the capitalist in the quest for capital accumulation.

It is in the interest of the capitalist or capitalists (nowadays, capitalists may, of course, consist of a number of shareholders, for example, rather than outright owners of businesses) to maximize profits and this entails (in order to create the greatest amount of new value) keeping workers' wages as low as is 'acceptable' or tolerated in any given country or historical period, without provoking effective strikes or other forms of resistance. Therefore, the capitalist mode of production is, in essence, a system of exploitation of one class (the working class) by another (the capitalist class).

Note

This appendix is reproduced from Cole, 2008d, pp. 24–25.

Conclusion

I began this book by discussing the origins of CRT in Critical Legal Studies (CLS). I went on to juxtapose the CRT concepts of 'white supremacy and "race" as primary' with Marxist analyses of social class and racialization. While recognizing that CRT has a number of strengths, I argued that Marxist analysis enhanced these strengths.

I then addressed multicultural education in the United States and United Kingdom respectively, and discussed antiracist responses (based on Marxism) in each country. Having made a number of references to neoliberal global capitalism and imperialism, I went to examine these issues conceptually and materially. I then did the same for Marxism and twenty-first-century socialism. Finally, I looked at some areas of agreement between Critical Race Theorists and Marxist's regarding classroom practice, before critiquing one specific Critical Race theorist's classroom pedagogies. I contrasted these with some suggestions based on Marxism.

Like Weberianism, post-structuralism, postmodernism, and transmodernism, CRT appears to me to be ultimately lacking in a direction for moving humankind forward progressively. As far as Weber is concerned, he believed that socialism would be even more rationalized, and bureaucratic than capitalism and thus more alienating. A common criticism of post-structuralism and postmodernism is that, in focusing on deconstruction, they have no solutions, while for transmodernist, Enrique Dussel, as noted in the Introduction to this volume, the solution is an 'ex nihilo utopia'.

CRT and Human Liberation

Darder and Torres (2004, p. 98) observe, in the CRT view of education: ' "racial" liberation [is] embraced as not only the primary but as the most significant objective of any emancipatory vision of education in the larger society' (ibid.). According to Krenshaw et al. (1995b, p. xiii) Critical Race Theorists also share 'an ethical commitment to human liberation' but 'often disagree among [themselves], over its specific direction' (ibid.). Thus often in CRT the solution is vague. To take an example, introducing their edited collection, *Critical Race Theory in Education*, Dixson and Rousseau (2006b) talk about 'the struggle' (pp. 2–3); 'a vision of hope for the future' (p. 3); 'social action toward liberation and the end of oppression' (p. 3); 'the broader goal of ending all forms of oppression' (p. 4); and 'the ultimate goal of CRT—social transformation' (p. 7). To take another example, Dixson and Rousseau (2006, pp. 2–3) argue that 'CRT scholars acknowledge the permanence of racism' but that this should lead to 'greater resolve in the struggle'. They also refer to a CRT focus on 'praxis', which incorporates 'a commitment not only to scholarship but also to social action toward liberation and the end of oppression' (p. 3). They talk of 'eliminating racial oppression as part of the broader goal of ending all forms of oppression' (p. 4), and state that the 'ultimate goal of CRT [is] social transformation'. However, no indication is given of what they are struggling towards, what liberation means to them, or what is envisioned by social transformation and the end of all forms of oppression.

Mills is somewhat clearer. As we saw in chapter 2 of this volume, for him (1997, p. 111) '[w]hite Marxism [is] predicated on colorless classes in struggle'. He argues that if socialism is to come then 'white supremacy/majoritarian domination' must be overthrown first in 'the struggle for social democracy'. Only after 'white supremacy' has been overthrown, and 'social democracy' established is the next stage—socialism—possible. This seems to be in line with Mills' argument that 'a non-white-supremacist capitalism is morally and politically preferable to…white-supremacist capitalism' (reiterated in Pateman and Mills, 2007, p. 31 and Mills, 2007, p. 243), something with which I would totally concur. However, given the massive advantages to capitalism of racialized capitalism, capitalism without racism (or sexism) is almost inconceivable.

Whether, in the light of the current 'credit crunch' (a euphemism for the inherent contradictions in capitalism) capitalist politicians globally will adopt long-term a more 'social democratic' as opposed to 'neoliberal' form (they have already adopted interventionist measures in the short-term) remains to be seen. Certainly a number of commentators are urging this (e.g., Elliott, 2008; Irvin, 2008). Whatever happens, it is Marxism, I believe, that provides the possibility of a viable equitable future. In chapter 7, I posited developments in South America, specifically Venezuela, as providing one possible future direction for twenty-first-century socialism. Though currently a capitalist state, with a government enacting social democratic measures, Chávez *is* promoting socialist values and forms of organization. In the barrios of Caracas, and everywhere else where the poor live, and the spark of socialism has been lit, people are not celebrating Max Weber or post-structuralism; they are not embracing postmodernism, transmodernism or Critical Race Theory (for these are largely academic pursuits). Instead they are engaging with the possibility of a *practical* democratic socialism, a socialism that is truly inclusive, with respect to 'race', but also with respect to gender, sexual orientation, disability, age and other forms of exploitation and oppression.

The Legacy of Martin Luther King Jr.

It is worth recalling that, at the beginning of this volume, I recounted that one of the people cited by Delgado and Stefancic (2001, p. 4) as being influential in the genesis of CRT was that tireless and irrepressible campaigner against racism, Martin Luther King Jr. At the time of writing (Summer 2008), it is the fortieth anniversary of King's assassination. King, a reformer, pacifist and Baptist minister rather than a revolutionary socialist (Martin, 2008a), is quite accurately known for his gradualism and his reformism.

However, it is significant that in the year preceding his death King became notably radicalized. Charles Steele, 2008 president of the Southern Christian Leadership Conference (SCLC) (King was the first president) has emphasized that, towards the end of his life, King had moved on from purely 'racial' issues, and that his final campaigns were focused on fighting poverty and on labor disputes (cited in Harris, 2008).[1] Steele believes that King, who came to Memphis in 1968 in support of striking workers (Harris, 2008), 'was killed [there] because he had started to focus on poor folks, regardless of their colour' (cited in ibid.). As Jerald Podair puts it, '[i]f you thought having a talk about race was difficult in America, then having one about class is even harder' (cited in ibid.). Paul Harris (2008) concludes that '40 years ago King tried to start that debate as well. A bullet cut short his ambitions' (Harris, 2008).[2]

The implications for the subject matter of this book are clear. As long as CRT centralizes 'race' rather than class, and as long as it voices no serious challenge to United States and world capitalism, it will be tolerated. As Roland Sheppard (2006, p. 7) notes, Martin Luther King had a different perspective at the time of his death to the 1963 'I have a dream' speech: 'he had begun to view the struggle for equality as an economic struggle and the capitalist economic system as the problem'. As King, who by 1967 believed that the total elimination of poverty was now a practical responsibility (Sheppard, 2006, p. 8), put it in a speech to the SCLC in August, 1967:

> We've got to begin to ask questions about the whole society. We are called upon to help the discouraged beggars in life's marketplace. But one day we must come to see that an edifice which produces beggars needs restructuring. It means that questions must be raised. 'Who owns this oil? ...Who owns the iron ore? ...Why is it that people have to pay water bills in a world that is two-thirds water?' (cited in Sheppard, 2006, p. 8)

However, perhaps Martin Luther King's most unequivocal declaration of a firm change of direction came earlier, in remarks to his staff at the SCLC on November 14, 1966. King proclaimed that the civil rights reforms of the early 1960s 'were at best surface changes' that were 'limited mainly to the Negro middle class'. He went on to add that demands must now be raised to abolish poverty (cited in Martin, 2008a):

> You can't talk about solving the economic problem of the Negro without talk-
> ing about billions of dollars. You can't talk about ending the slums without
> first saying profit must be taken out of slums. You're really tampering and
> getting on dangerous ground because you are messing with folk then. You are
> messing with captains of industry.... Now this means that we are treading
> in difficult water, because it really means that we are saying that something
> is wrong...with capitalism.... There must be a better distribution of wealth
> and maybe America must move toward a democratic socialism. (cited in The
> Democratic Socialists of Central Ohio, 2008)[3]

Classism or Marxism and Democratic Socialism?

David Gillborn (2008, p. 13) may be right when he asserts that 'the best critical race theorists are passionate about...classism'. But while challenging the
oppression of people that is based on their social class (classism) is extremely
important, and is championed by Marxists, the fundamental point is to also
challenge the *exploitation* of workers at the point of production, for therein
lies the economic relationship that sustains and nurtures the capitalist system. While I am critical of CRT, I would like to reiterate that the purpose
of this book is to not to divide, but to unite. My intention has not been to
question the ideological or political integrity of Critical Race Theorists, but
to open up comradely discussion in the light of the entry of CRT into British
Academia. In chapter 5 of this volume, I discussed David Gillborn's reluctance to engage in debate with Marxists. However, there seem to be some
contradictions in his position, because he has also argued that 'the best way
ahead may simply be to make use of analytical tools as and when they seem
most revealing' (Gillborn, 2008, p. 38). This is followed up by the assertion
that Marxists (presumably) will not be amenable to this. He states: 'this will
not satisfy people who seek to fetishize a single concept or theory above all
else' (ibid.) He then goes on to emphasize what he sees as central tenets of
CRT. He quotes Kimberlé Crenshaw (1995, p. 377) as follows:

> Through an awareness of intersectionality, we can better acknowledge and
> ground the differences among us and negotiate the means by which these dif-
> ferences will find expression in constructing group politics (cited in Gillborn,
> 2008, p. 38)

To make matters even more confusing, Gillborn (ibid.) then cites David
Stovall (2006, p. 257) as stating that '[a]rguing across conference tables
is useless', that our work must be done 'on the frontline with communities committed to change' and that 'neither race nor class exists as static
phenomena'.

For Marxists, there is a need for both arguing across conference tables and
working on the frontline with communities committed to change. Dare I
urge Gillborn, in a comradely way, to reconsider this reluctance to talk with

Marxists, and, in so doing, perhaps address himself to some of the strengths of Marxism?

As noted in the Introduction to this volume, it was Max Weber who is credited as being the first prominent *sociologist* to dispute Marx's arguments in a serious way. Since then, there have been many other academics who have sought ways to challenge Marx and Marxism, Critical Race Theorists, being among the most recent. There will no doubt be many others. Marxists will continue to meet these challenges. Marxism is not, as some would have it, a moribund set of beliefs and practice. On the contrary, as noted in chapter 5 of this volume, Jean-Paul Sartre (1960) has described Marxism as a 'living philosophy'. To Sartre's observation, Crystal Bartolovich (2002, p. 20) has added, Marxism is not 'simply a discourse nor a body of (academic) knowl-edge' but a living project. As I have stressed, the Bolivarian Misiones are classic examples of social democracy rather than socialism. It is important to recognize that no one can foresee what direction the Bolivarian Revolution will take. Like other revolutions, it may be defeated or it could be hijacked. However, the Bolivarian Revolution is firmly placed in the dialectics of socialist struggle, and its full effect on Cuba, and the emerging struggles in other parts of South America and possibly the rest of the world are yet to be seen.

Whether or not the Bolivarian Republic of Venezuela remains a capital-ist country, or proves to be a concrete example of an attempt to nurture the living project of Marxism remains to be seen. It will certainly not be the last such attempt. The struggle against capital and empire was important historically, continues unabated in the present, and will be mounted against any empire of the future. As Hugo Chávez put it in a speech to the United Nations (cited in Ali, 2008, p. 293):

> We reaffirm our infinite faith in humankind. We are thirsty for peace and justice in order to survive as a species. Simón Bolívar, founding father of our country and guide to our revolution swore to never allow his hands to be idle or his soul rest until he had broken the shackles which bound us to the empire. Now is the time not to allow our hands to be idle or our souls to rest until we save humanity

A Realignment of CLS and CRT Informed by Marxism?

In chapter 1, I outlined the central tenets of Critical Legal Studies, pointing out its location in economic and social class analyses, and indicated its essen-tially socialist credentials. I also recounted the break from CLS of CRT, pri-marily because of this emphasis, but also because a number of people of color needed a voice, hitherto not available, to bring 'race' firmly to the center of analysis. In the twenty-first century, CRT is now firmly established, while CLS has essentially disbanded. In chapter 2, I pointed out that Richard Delgado had argued for the importance of social class some five years ago. Perhaps a re-alignment of CLS and CRT might be worth considering as a fruitful partnership in the important, indeed crucial, tasks that lie ahead for all progressive people, a partnership which, as I stressed in chapter 2, would need to centralize capitalism and capitalist social relations. *I would like to appeal to Critical Race Theorists and Critical Legal Studies scholars to join me in productive antiracist dialogue in order to see if we can agree how to move theory and action (praxis) forward.*

On a more societal level, can we return to Duncan Kennedy's strategy, outlined in chapter 1, of building an *inclusive* mass movement for the radical transformation of American Society? And indeed the world.

Postscript

On November 5, 2008, America elected its first black president. As argued in chapters 3 and 6 of this volume, this momentous event will not, in itself, move the United States forward in a progressive direction. Indeed, as noted in chapter 6, there are no significant indications of a major change in policy.

The *London Evening Standard* billboard at Kings Cross Station announced on the day of the election victory, 'Obama Mania: Shares Surge', thus vindicating instantaneously interest convergence—but not in CRT parlance between people of color and white people per se—more a shared interest in the election of a black Democrat between people of color and national (and international) capital. As Anindya Bhattacharyya (2008, p. 7) points out, while 73 percent of the poorest households (with an annual income of less than $15,000) voted for Barack Obama, so did 52 percent of the richest households (those earning over $200,000 a year). Top-down corporate mobilization for Obama, as Bhattacharyya (ibid.) points out, meant that by 'mid-October [Obama] had raised a massive war chest of $640 million and spent $250 million on TV advertising', while McCain's October budget was $47 million (ibid.). Moreover, as Alex Lantier (2008) points out, Obama's transition team is co-chaired by a Chicago real-estate magnate, and a former chief of staff for President Bill Clinton and head of the Podesta Group, a Washington lobbying firm. The team employs 450 people and has a budget of $12 million. The co-chairs of the U.S. Treasury review team are an investment banker, and the chief operating officer of a Washington lobbying firm Stonebridge International LLC, while the co-chairs of the State Department review team are a former top lobbyist for U.S. mortgage giant Fannie Mae and a top employee at the Albright Group, an international lobbying firm founded by Clinton administration secretary of state Madeleine Albright.

The presence of this corporate capitalist machine behind Obama is not to deny the symbolic importance of Obama's victory. The unmitigated joy and pride of the people of color of America and elsewhere transmitted globally surely brought tears to the eyes of even the most cynical antiracist. However, in CRT terms, as I argued on p. 57 of this volume, the election of Obama may well be a major instance of a contradiction-closing case—it might become more difficult to uphold charges of racism in that deeply racist society: how can America now be racist, when a black child can become president? Historian

Simon Schama told the BBC that this election 'wipes away America's original sin' (Weaver, 2008b), while five days after Obama's election, a black reader, Winston Drake, wrote to a free UK-based newspaper (*Metro*, November 10, 2008, p. 18): '[t]here is now absolutely no reason for black people to complain they are mistreated racially'. The election of a black president will not end racism in the United States. If the arguments of this book are right, if racialization occurs in relation to varying requirements of the capitalist mode of production, then racism will exist as long as capitalism does. Capitalism's abolition is a necessary if not a sufficient step in doing away with it (San Juan, Jr., 2008).

On the same day as Drake's letter appeared in the *Metro*, *Guardian* columnist Gary Younge (2008, p. 27) provided some more realistic grounds for optimism, noting, that 'tens of thousands of volunteers' worked for Obama (The *Guardian*, p. 27), and that 'the most interesting thing about Obama has always been his base...the black, the young, the Latino and the poor' (ibid.) (as indicated by the percentages above). Younge goes on, '[n]ow we'll see whether this electoral base has the will and wherewithal to transform itself into a potential movement that might both support and challenge him' (ibid.). As one of the new President's volunteers, Steve Thompson, reminded us 'Obama kept saying, it's not about me. It's about you' (cited in ibid.). Younge (ibid.) concludes

> The first act is over. The question now is who will write act two? The protagonists should not cede the stage, lest the powerful shape the narrative.

Indeed. For Marxists, these temporary celebrations must be translated into calls for the end of the system that sustains and promotes racism, imperialist wars, class exploitation, and other injustices. Only with the demise of capitalism can celebration be long lasting. Only then can symbol become real and material. In addition to the fact that the massively discredited Bush is no longer of use to capitalists, the election of a more 'liberal' and apparently more pro-worker (but in reality just as pro-capitalist and imperialist—see pp. 57–59 of this volume) president makes it easier to manage the workers as a class, to control their organizations, and to make sure it is not the rich (other than perhaps a symbolic handful) who pay for the current financial crisis (Dave Hill's comments on this postscript). However, as Anthony Green has pointed out (his comments on this postscript), there are some grounds for optimism, given that the election victory has come at a time when the inherent instability of capitalism is there for all to see and to experience. The inevitability of 'boom and slump' in the capitalist system, paradigmatic for Marxists, is firmly on the international agenda, and can provide grounds for optimism of the will and cause for a renewal of energy for all progressive people. Moreover, it provides space for educators to nail the lie that 'there is no alternative' (TINA) to neoliberal capitalism; it provides a lacuna to debate the real alternative to capitalism and imperialism. A debate about Marxism and socialism is not only more possible, it is also more necessary than ever.

Notes

Introduction

1. The city had very few black inhabitants in the early 1950s. The estimated 5,000 slaves who had been kept as servants in the eighteenth century had long since merged into the general population (Dresser, 1989, p. 310).
2. In the UK context, 'Asian' has a similar meaning to 'South Asian' in the U.S. context.
3. During an impromptu feedback to the head at the end of the day on how things went, I happened to mention that, after we had done some Math and some English, we then played some reggae. The response of the head teacher of this predominantly African-Caribbean multicultural school was, 'oh, what's that?'
4. Perhaps such a challenge began with the sociologist Max Weber. Although Albert Salomon's famous observation that Weber was involved in a debate 'with the ghost of Marx' (Salomon, 1935) may be somewhat overstated, ever since Weber (c. 1915) made a number of criticisms of Marx and Marxism the intellectual struggle against Marxist ideas has been at the forefront of academic writing. Weber suggested that social class might not be solely related to the mode of production; that political power does not necessarily derive from economic power; and that status as well as class might form the basis of the formation of social groups. Subsequent attempts to challenge Marxist ideas have ranged from the post-structuralist writings of Michel Foucault who believed that power is diffuse rather than related to the means of production, and of Jacques Derrida who stressed the need for the deconstruction of all dominant discourses; through the postmodernism of Jean-François Lyotard who was incredulous of all grand narratives, of Jean Baudrillard who argued that binary oppositions (such as the ruling and working classes) had collapsed; to the transmodernism of Enrique Dussel who has reworked the limited postmodern notion of multivocality (where all voices have equal validity) in favor of the prioritizing of the voices of 'suffering Others'. Although renowned for his scholarly, thorough, ongoing and original reading of Marx, and for his numerous publications on various aspects of Marxism, Dussel, nevertheless advocates an 'ex nihilo utopia' (a utopia created from scratch) rather than socialism (see Cole, 2008d, Chapter 6). Elsewhere (Cole, 2008d), I have defended Marxism against these various challenges, but have also acknowledged some of the insights of these diverse theories.

1   Critical Race Theory: Origins and Varieties

1. Analogic Reasoning, for the founder of transmodernism Enrique Dussel, is reasoning outside of the class struggle, whence the new utopia will be born (for a discussion of Analogic Reasoning, see Cole, 2008d, pp. 72–74). This is particularly problematic for Marxists since, for them, the new society will be born as a result of the class struggle. Marxists believe that social change comes about through struggle as part of a *dialectical* process via thesis, antithesis, and synthesis. The warring classes are always the products of the respective modes of production, of the *economic* condition of their time. Thus slaves were in struggle with their masters in the historical epoch of ancient slavery. This gave way to feudalism (synthesis). Feudal serfs were in struggle with their lords in times of feudalism, and workers are in struggle with capitalists in the era of capitalism. This struggle is exacerbated in the mode of neoliberal global capitalism and modern imperialism (see chapter 6 of this volume). Marxists believe, given the right circumstances (workers becoming class conscious and seeing through the interpellation process—discussed at the end of this chapter—and realizing that they *are* in struggle with capitalists), this can lead to the synthesis of socialism.

2. Analectic Interaction involves listening to the voices of 'suffering Others' and interacting democratically with them as a step towards liberation (see Cole, 2008d, pp. 74–75).

3. A distinction needs to be made between the Marxist usage of the term 'working class' and sociological meanings of the term. For Marxists, the working class consists of *all* those who need to sell their labor power to survive rather than living off the labor power of others (see the appendix to chapter 8 of this volume). Sociologists such as Weber use the term to describe those and their children in lower status occupations with lower earnings. Both definitions have their advantages and both are used throughout this volume.

4. Legal realism is a family of theories about the nature of law developed in the first half of the twentieth century in the United States and Scandinavia. The essential tenet of legal realism is that all law is made by human beings and thus is subject to human foibles, frailties and imperfections. Thus, as Woodard (1986, p. 1) argues, for legal realists, policy choices 'should be informed by the best knowledge, legal or extralegal, and not based soley on the the artificial authority of earlier cases'. In the case of the United States, it disavowed mechanical jurisprudence in favor of insights from social science and politics and utilized policy judgment (Delgado and Stefancic, 2001, p. 150).

5. The term 'liberalism' tends, in popular (and academic) parlance, to be used differently in the United Kingdom and the United States. Traditionally in the United Kingdom, it has been used (in popular discourse by Marxists and other Left radicals) to describe 'middle of the road' politics. In the United States, it has often been used in everyday parlance to describe those who are viewed to be on the Left. The 'Marxism-aware' CLS writers discussed in this chapter, including Tushnet, tend to use 'liberalism' to describe 'middle of the road' politics, as do similarly Marxism-aware Critical Race Theorists.

6. Unger (1986) held similar views, arguing for an alliance between disaffected elites and the downtrodden, and urging CLS lawyers to wage a strategic campaign of 'constructive dissidence' (cited in Woodard, 1986, p. 3).

7. Reading the twenty or so leading law review articles on civil rights in the early 1980s, Delgado ([1984] [1995], p. 46) discovered that the authors were

all white males, and that those they cited were also white males. Given that there were at the time about one hundred black, twenty-five Hispanic, and ten Native American law professors teaching in U.S. law schools, many of whom were writing about civil rights and related issues (pp. 46–47), Delgado (p. 47) concludes that much of their scholarship 'seems to have been consigned to oblivion'. He goes to argue that a number of the white males' 'inner-circle' articles showed a lack of awareness of basic facts of how minority persons lived or how they viewed the world (p. 49). Delgado also found some of their arguments naïve and 'hardly worthy of serious consideration'. Most troubling for Delagado was that such articles are not confined to academia, but are cited by courts, and their ideas are 'read and discussed by legislators, political scientists, and their own students' (p. 51). Arguing in a similar vein a couple of years later, Harlon L. Dalton ([1987] [1995]) suggested that 'the quite distinct social circumstances of white males has led to a "rights critique" that is oblivious to, and potentially disruptive of, the interests of people of color' (ibid.). Delgado (1992) followed up his 1984 study, and found that, while some of the old 'inner circle' had retired, they had been replaced by a new 'inner circle'. Unfortunately, although minority writers were writing in top journals, not much had changed and these new minority writers had not been fully integrated into the conversations of the new 'inner circle' (Isaksen, 2000, p. 698).

8. As Du Bois (1915, p. 139) put it some twelve years later, too many have accepted 'that tacit but clear modern philosophy which assigns to the white race alone the hegemony of the world and assumes that other races...will either be content to serve the interests of the whites or die out before their all-conquering march'.

9. Although the students at Harvard may have been the *original* impetus for CRT, as Richard Delgado (personal correspondence, 2008) notes, 'when they graduated, the momentum at Harvard died. The school purged Derrick Bell...and is today more conservative than ever'.

10. 'Racialism' in the UK context at least, is an old-fashioned term for 'racism'.

11. The article by Kennedy (1982) discussed in the chapter is a good example of a nondeterminist or anti-'vulgar' Marxist approach. At one point in the article, Kennedy specifically states, 'law cannot be usefully understood, by someone who has to deal with it in all its complexity, as "superstructural"' (Kennedy, 1982, p. 599).

12. Althusser (1971, p. 174) uses this formulation to describe the way in which ruling class ideology undermines the class consciousness of the working class. For Althusser, the interpellation of subjects—the hailing of concrete individuals as concrete subjects—'Hey, you there!' (p. 175) involves a four-fold process: (1) the interpellation of 'individuals' as subjects; (2) their subjection to the Subject; (3) the mutual recognition of subjects and Subject, the subjects' recognition of each other, and finally the subject's recognition of himself; (4) the absolute guarantee that everything really is so, and that on condition that the subjects recognize what they are and behave accordingly, everything will be all right: Amen—*So be it*' (p. 182). Althusser explains:

> Caught in this quadruple system of interpellation as subjects, of subjection to the Subject, of universal recognition and of absolute guarantee, the subjects 'work', they 'work by themselves' in the vast majority of cases, with the exception of the 'bad subjects' who on occasion provoke the intervention of one of the detachments of the (repressive) State apparatus.

But the vast majority of (good) subjects work all right 'all by themselves', i.e., by ideology (whose concrete forms are realized in the Ideological State Apparatuses). They are inserted into practices governed by the rituals of the ISAs. They 'recognize' the existing state of affairs...that 'it really is true that it is so and not otherwise', and that they must be obedient.

Subjects recognize that 'the hail' was really addressed to them, and not someone else (p. 175) and respond accordingly: 'Yes, that's how it is, that's really true!' (p. 140). Their subjection is thus freely accepted (p. 183). Thus when confronted with the 'inevitability' of neoliberal global capitalism, or TINA ('there is no alternative'), the response is 'That's obvious! That's right! That's true!' (p. 173). There is no point, therefore, to even consider alternative ways of running the world, such as democratic socialism, or even social democracy. (Of course, we must not forget that because of their material conditions of existence, participation in struggle etc, the 'bad subjects' might respond differently: 'that's most dubious! That's wrong! That's a lie. Things can be different'.) (My thanks to Susi Knopf for alerting me to the connection between Delgado, Ladson-Billings and Tate, and Althusser. Althusser's theory of Repressive State Apparatuses [RSAs] and Ideological State Apparatuses [ISAs] is discussed briefly in chapter 3 of this volume, and developed at length in Hill 1989, 2001b, 2004, 2005b. In the context of the United Kingdom and Venezuela, it is developed in Cole, 2008e, 2008g, 2008h.) My use of selected Althusserian concepts that I find illuminating—the power of capitalist structures to keep capitalism on course—does not signal agreement with the work of Althusser as a whole. A structuralist Marxist, Althusser is widely perceived to be deterministic, in denying the power of 'human agency'—the ability of people to successfully struggle to change things. Althusser's structuralist Marxism is often contrasted with the humanist Marxism of Antonio Gramsci. Gramsci famously called for continued determination, even in the direst of circumstances, in the belief that resilience will result in meaningful change even in the face of adversity (Gramsci, 1921).

2 White Supremacy and Racism; Social Class and Racialization

1. This chapter draws on and develops Cole, 2009a. For Mills' response to Cole, 2009a, see Mills, forthcoming, 2009. For my reply to Mills, see Cole, 2009d.
2. Although I am most critical of 'white supremacy' as will become clear in this chapter, I must acknowledge that it can play a useful role as 'shock factor'. For example, a black colleague once described how he got the staff of his school to take racism seriously by announcing at a staff meeting that the school was a white supremacist school.
3. Pateman (1988) argues that lying beneath the myth of the idealized contract, as described by Hobbes, Locke, and Rousseau, is a more fundamental contract, the agreement by men to dominate and control women. Men's relationships of power to one another change, but women's relationship to men's power does not. Patriarchal control of women, according to Pateman, is found in at least three contemporary contracts: the marriage contract, the prostitution contract, and the contract for surrogate motherhood. For Pateman, contract is not the path to freedom and equality. Rather, it is one means, perhaps the most fundamental means, by which patriarchy is upheld. Mills and Patemen

have since written a joint book (Pateman and Mills, 2007) that develops and defends the arguments in Pateman (1988) and Mills (1997) respectively. I was present at the launch of Pateman and Mills' (2007) book late in 2007 in Cardiff, a meeting that underlined my positive feelings about their integrity. The following critique of Charles Mills should be read as comradely criticism in the pursuit of our joint and burning desire to see a world free of racism.

4. Another problem is that McIntosh's (1988, pp. 5–9) list is actually composed of forty-six privileges. Gillborn is not alone is misinterpreting the work of Peggy McIntosh. There are tens of thousands of unauthorized versions of her work, most of it inaccurate, posted on the Internet.

5. Color-coded racism in the United Kingdom is indicative of a capitalist society, with a continuity of racism from its imperialist past, rather than a white supremacist society. In the twenty-first century, such racism continues unabated with respect to the descendents of the people of the former colonies, both black people, and Asian people. Elsewhere, I have documented color-coded racism and its relationship with developments in capitalism and imperialism from the days of Empire, through the post-war period, up to the present, both in general terms (e.g., Cole and Virdee, 2006), and with respect to education (e.g., Cole, 2004a; Cole and Blair, 2006).

6. In Brighton, UK, where I live, I recently encountered anti-Irish racism from an interesting and somewhat contradictory source. Many, if not most, of the white Brighton taxi drivers are rampant racists (and sexists), and often assume that I will be too. When the driver asked me where I was going, I explained that I was going to a conference to discuss human rights. At that point, he remarked that human rights were close to his heart because his father was a Romany. An antiracist white taxi driver at last I thought! After he had enlightened me with tales of anti-Romany racism, he concluded. 'It's the Irish Romanies I can't stand, they're dirty; they don't wash'. 'Not all of them, surely?', I interjected. 'Yes', he replied. 'Every single one. They don't wash'. The taxi drivers of color, by contrast, are wary of discussing racism, perhaps because, like the white taxi drivers, they too presume that I am racist.

7. For a Marxist analysis of migrant labor in contemporary Europe, see Dale and Cole, 1999b. Racialization directed at the working classes of Europe is not a new phenomenon. The British State has applied it from the fifteenth century onward, to the Dutch, the Huguenots, Eastern European Jews, and others.

8. Xeno-racism is a problem in Lincoln, UK, where I work. A short while ago, the manager of my local pub had to make a citizen's arrest of someone who, hearing a fellow drinker speaking Slovak, shouted abuse, asserting that Eastern Europeans were taking locals' jobs. The abuser then hit the Slovakian over the head with a chair.

9. Mills (1997, pp. 126–127) underlines this point, when he makes a distinction between 'whiteness as phenotype/genealogy and Whiteness as a political commitment to white supremacy'. Indeed, Mills (p. 127) argues that 'whiteness is not really a color at all, but a set of power relations'. My response to this would be, 'then why not use racialized capitalism' as a descriptor? As long as 'white supremacy' is maintained, it will continue to have explicit for most people (and implicit for others better versed in the intricacies of CRT) connections to skin color. Mills (forthcoming, 2009) has argued that the concept of 'racialized capitalism' obfuscates the realities of white working class agency and historic complicity in various forms of racism, 'imperialism, colonialism, genocide,

apartheid, the "color bar" and Jim Crow'. I have two responses to this. First, there is no doubt that there are large amounts of evidence for such complicity. From a Marxist perspective, while there are material and psychological benefits (as Mills, forthcoming, 2009 points out) to the white working class in acquiescing to racist structures and practices, the fact that they do acquiesce is best explained by 'false consciousness' and, as is argued throughout this volume, by the success of the Repressive and Ideological State Apparatuses (Althusser, 1991) in interpellating subjects in the interests of capitalism. Second, it must be pointed out that there is also evidence of white working class resistance to racism. A good UK example is the formation of a current of white working-class antiracism in Britain in the 1970s and 1980s (for details, see Cole and Virdee, 2006, pp. 46–49).

10. Elsewhere (Cole, 2008d, p. 153), I noted how this was underscored when, in a discussion in a café with a Marxist friend, I mentioned the organization, *Race Traitor,* and he told me to 'hush' in case we were misunderstood. A related incident happened recently on a train. I travel frequently between Brighton and Lincoln, and, although I meticulously recycle paper at home, I sometimes leave ongoing but discarded drafts of the manuscript of this volume in the waste bins of trains. Having put a sheet of paper in such a bin, I had to hastily retrieve it, when I saw the words, 'proof of white supremacy' clearly visible in the over-spilling bin.

11. Jeff Nixon (2008, p. 81) has summed up the scope of the Act:
The General Duty section of the Act (Home Office, 2000) has three parts:
  - eliminate unlawful racial discrimination
  - promote equality of opportunity
  - promote good relations between people from different racial backgrounds.
In relation to schools this means that policies and statements covering admissions, assessments, raising attainment levels, curriculum matters, discipline, guidance and support, and staff selection and recruitment, should all have elements which address the three parts of the General Duty. Schools have to bear in mind that the size of the minority ethnic population does not matter; racial equality is important even when there are no minority ethnic pupils or staff in a school or local community. Schools must have a written statement of policy for promoting 'race' equality and arrangements for assessing the impact of policies on pupils, staff and parents/carers, and a system of monitoring the operation of the policies paying particular attention to the levels of attainment of pupils from minority ethnic groups. The Race Equality Policy must be a clearly identifiable and easily available part of the school's policy on equal opportunities or the policy on inclusion; there should also be a clear link between the policy and the school's action plan or the School Improvement Plan. All racist incidents, whoever perceives them to be racist, irrespective of whether they are on the receiving end, must be investigated and reported to the governing body every school term.

Unfortunately, the efficacy of the Race Relations (Amendment) Act (2000) has been undermined in Britain in recent years by a new stress on community cohesion (see chapter 4).

12. That this is the case is explained succinctly by Marxist geneticists Steven Rose and Hilary Rose (2005; see also Darder and Torres, 2004, pp. 1–12, 25–34).

As they note, in 1972, the evolutionary geneticist Richard Lewontin pointed out that 85% of human genetic diversity occurred *within* rather than *between* populations, and only 6%–10% of diversity is associated with the broadly defined 'races' (Rose and Rose, 2005). As Rose and Rose explain most of this difference is accounted for by the readily visible genetic variation of skin color, hair form, and so on. The everyday business of seeing and acknowledging such difference is not the same as the project of genetics. For genetics and, more importantly, for the prospect of treating genetic diseases, the difference is important, since humans differ in their susceptibility to particular diseases, and genetics can have something to say about this. However, beyond medicine, the use of the invocation of 'race' is increasingly suspect. There has been a growing debate among geneticists about the utility of the term and in Autumn 2004; an entire issue of the influential journal *Nature Reviews Genetics* was devoted to it. The geneticists agreed with most biological anthropologists that for human biology the term 'race' was an unhelpful leftover. Rose and Rose conclude that '[w]hatever arbitrary boundaries one places on any population group for the purposes of genetic research, they do not match those of conventionally defined races'. For example, the DNA of native Britons contains traces of multiple waves of occupiers and migrants. 'Race', as a scientific concept, Rose and Rose conclude, 'is well past its sell-by date' (2005). For these reasons, I would argue that 'race' should be put in inverted commas whenever one needs to refer to it.

13. To this assertion, I would ask by whom? Mills' context is academic writing, so I'll bypass the nearly six million members of the Venezuelan United Socialist Party (see chapter 7 of this volume), and the fact that *The Communist Manifesto* remains a bestseller, and just note, from my own discipline, *Educational Theory*: the readers of the *Journal for Critical Education Policy Studies* (http://www.jceps.com/) that has had over 300,000 hits since its inception in 2003; the 155 members of *Marxian Analysis of Schools, Society, and Education Special Interest Group (MASSES)* SIG (Special Interest Group) of AERA (the American Education Research Association); and the two major series on Marxism and Education from both Routledge and Palgrave Macmillan. Mills (forthcoming, 2009) is right in noting that the fact that his book, *The Racial Contract* sold 20,000 copies represents, in academic terms, a 'success'. Using the same criterion, it is difficult, from the figures above, to draw the conclusion that (academic) Marxism is dead. I suspect that some of Mills' 'thousands of American college students' exposed to CRT, as a result of Mills' book (Mills, forthcoming, 2009) have not been unaffected by the growth of Marxist scholarship.

14. I would personally prefer to use a nondisablist term, perhaps 'color-neutral'? If there are any doubts in the reader's mind about the negative use of 'blind' being offensive, then it is probable that the reader has not worked with a blind colleague or taught a blind student. The same applies with the term, 'deaf'. The common expressions, 'are you blind' or 'are you deaf?' take on new (offensive) meanings when working with or teaching blind, sight-impaired, deaf or hearing impaired students.

15. As we shall see in Chapter 5, in Pateman and Mills, 2007 (p. 32), Mills has called for 'non-white-supremacist capitalism', something that he states he has 'been arguing in more recent work' (ibid.).

16. Youdell et al. (2008) explain how the capitalist class gain from (xeno-) racism and (xeno-) racialization:

    If one group of society is badly off and marginalised, why would this be the fault of another group that is badly off and marginalised? It doesn't stand up to inspection. But it is an old refrain that pits one disadvantaged group against another and leaves the privileged and the powerful to stand by and watch, disapproving and unscathed. (ibid.)

    Instead, they suggest, we should be looking at the long-term impact of the irreconcilability of deindustrialisation and consumerism for the (once) white working class, and interrogating the global, national, regional and local economic relations and policy priorities that have created lack of social mobility, unemployment and disenfranchisement for both white and minority ethnic working-class people.

17. An exception was the suggestion of one contributor that we should use 'Englishness' rather than racism. His argument was that 'Englishness' helps to explain the contemporaneous incorporation of previously racialized groups. My response would be that the connection between 'racism' and 'nationalism' implicit in 'Englishness' may be particularly close in the English context. Miles has, in fact, argued that while it cannot be argued that there is a necessary relation between 'race' and 'nation', such articulation is particularly strong in the case of England (Miles, 1989, pp. 89–90). As he puts it, 'English nationalism is particularly dependent on and constructed by an idea of "race", with the result that English nationalism encapsulates racism' and that 'the ideas of "race" and "nation", as in a kaleidoscope merge into one another in varying patterns, each simultaneously highlighting and obscuring the other' (Miles, 1989). Thus while the concept of 'Englishness' may explain a form of *racism* (excluding Others not considered to be 'English') at a specific juncture—e.g., what was perhaps occurring in England at the time of the conference—'racism' is a more useful *general* term to describe discourse, actions, processes, and practices, both historically and contemporaneously.

18. The thrust of Herrnstein and Murray's (1994) arguments in *The Bell Curve* is an unabashed defense of social inequality, attributing wealth and poverty to superior versus inferior genetically determined intellectual abilities. The political conclusion of *The Bell Curve* is a rejection of all policies aimed at ameliorating social injustice and furthering democratic values.

19. Ellis, erstwhile lecturer in Russian and Slavonic Studies at Leeds University, told the Student newspaper that he supported the theory developed by Herrnstein and Murray that white people are more intelligent than black people. He also said that women did not have the same intellectual capacity as men (Taylor, 2006). According to Matthew Taylor, Ellis first came to prominence in 2000 'when he traveled to the US to speak at the American Renaissance conference, an event described by anti-fascist campaigners as a three-day rally bringing together the scientific racism movement'. Apparently the event attracts organizations such as the Ku Klux Klan. In one of his books, Ellis states that the fascist British National Party (BNP) 'is the only party in Britain that has consistently attacked the scandalously high levels of legal and illegal immigration'.

20. Steve Fenton (2003, p. 164) has described making a distinction between the (ethnic) majority, an almost unspoken 'us', and members of minority ethnic

communities as 'ethnic majoritarian thinking'. It is perhaps epitomized in the title of the (1981) Rampton Report: *West Indian Children in Our Schools.* The distinction is underlined by the fact, for example that British Muslims have to substantiate their allegiance to Britain. After the Forest Gate terror raid (where the police raided the home of two innocent Muslim brothers, one of whom was shot, though not fatally; see Cole and Maisuria, 2008), the media highlighted the fact that the brothers stated that they were 'born and bred' East Londoners and they 'loved Britain' (Getty, 2006, p. 5).

21. As Hill (2001a, p. 8) has pointed out, the influence of ideology can be overwhelming. He cites Terry Eagleton (1991, p. xiii) who has written: '[w]hat persuades men and women to mistake each other form time to time for gods or vermin is ideology'. This observation is particularly applicable to the concept of racialization.

22. In adopting Miles' definition of racialization, I should make it clear that there are a number of non-Marxist applications of the concept of racialization. Indeed, the concept is a contested term, which is widely used and differently interpreted (for an analysis, see Murji and Solomos, 2005).

23. I recognize the problematic nature of the terms 'asylum-seeker' and 'refugee'. They form part of a 'discourse of derision' (Ball, 1990, p. 18) in the media, and in the pronouncements of certain politicians. 'Forced migrants' (Rutter, 2006) might be a more appropriate term.

24. For those migrants from the EU who do not have white skins, the (traditional) concept of racialization would, of course, apply.

3   The Strengths of CRT

1. The online version has this sentence incorrectly translated as 'Other than this neither Marx nor I have ever asserted'. My thanks to Stathis Kouvelakis for pointing this out to me.

2. Here I am using 'white supremacy' in the conventional sense, where it refers to regimes such as apartheid South Africa and Nazi Germany, as well the Southern states of the United States before 'civil rights'.

3. 'Creole' is an adaptation of the Castillian-Spanish colonial word, criolla. In the U.S. context, the term 'Creole' refers to a unique group of people who reside in Louisiana, particularly New Orleans. The first Creoles were descendants from Senegambia in Africa, who were brought to Louisiana as slaves in the eighteenth century. Creoles are of mixed African and French and/or Spanish heritage (DeCuir-Gunby, 2006, p. 96).

4. DeCuir-Gunby (2006, p. 107) argues that the case of Josephine DeCuir 'evokes many questions about the present-day role of mulitiracialism', making reference to the 2000 U.S. Census that for the first time gave individuals the opportunity to 'identify multiracially' (p. 108). Noting that on the surface this option seems positive, beneath the surface she suggests, 'listing all racial groups, especially white, is an attempt to access the property rights of whiteness' (p. 108), and that the 'multiracial label' has been most used by whites 'in an attempt to transfer the property rights of whiteness to their multiracial children'. While the motives of people using 'multiracial labels' are interesting issues in the context of a racialized capitalist society, I do not think that directly relating the case of Josephine DeCuir to such issues is particularly illuminating. What I have attempted in this section of the chapter is to assess CRT's efficacy in explaining *segregation* and *white supremacy* in

their historical contexts. The case of Josephine DeCuir performs this function well.

5. In Marxist theory, the capitalist state is more than mere government, and includes a range of institutions, including the hierarchy of the police, of the armed forces, the courts and so on. I am thinking here of national states. There are also local states, as, for example, the individual states of the 'United States of America', each of which also encompasses a range of institutions.

6. This is not surprising given the fact that, in her writing in general, Ladon-Billings shows more sympathy to Marxist theory and concepts than do many other Critical Race Theorists.

7. Murray's interpretation is most interesting. First, it implies that media images are somehow objective; second, in a most racist and sexist way, it juxtaposes black men and black women—the former, criminals, the latter, inept.

8. I put 'white' in brackets because it is possible to imagine a state, in which the ruling faction is not white, that might accommodate the interests of another nonwhite minority when it converges with the interests of the nonwhite state. Indeed, an example of this might be President Thabo Mbeki of South Africa's delayed response to the violent racism meted out in South Africa from black South Africans to black Zimbabwean immigrants (see chapter 5 for a discussion). Mbeki was silent for two weeks before eventually breaking his virtual silence, *following international coverage of these attacks* ('left more than 50 people dead and tens of thousands fleeing their homes') to denounce them as an 'absolute disgrace' (McGreal, 2008a).

9. It is also possible to imagine this scenario taking place when none of the participants are people of color. We might just substitute a metropolitan area offering school places to white working class students in predominantly white middle-class schools, and enticing middle-class students to attend city schools (I am using class in a sociological rather than Marxist sense here: in terms of occupation and status rather than relationship to the mode of production).

10. Delgado's addition of 'or the poor' is significant and, one would assume, relates to the aforementioned fact that Delgado favors a 'return to class' within CRT.

11. Gillborn (2008, p. 133) is careful to stress his awareness that his analysis could be seen as disrespectful to the Lawrence family' ongoing battle for justice, and of victories won along the way. He underlines that this is neither his intent, nor, he hopes the outcome of his analysis, and lists a number of such victories (pp. 133–134). 'The Lawrence Inquiry', he notes, 'has delivered considerable advances and holds out the possibility of further progress, but it is a start not an end' (p. 135).

12. Blair's and Brown's policies are indicative of what Gillborn (2008, pp. 81–86) refers to as 'aggressive majoritarianism' (see chapter 5 of this volume for a discussion).

13. Similar hypocrisy was apparent in the *Sun* that, as Gillborn (2008, p. 86) points out, transposed its usual anti-Muslim sentiment into the more acceptable language of gender equity ('face veil stops girls learning'), while on the opposite page it featured topless 'Nikkala 24, from Middlesex'.

14. Louis Althusser (1971) differentiates between what he calls the *Repressive State Apparatuses* (*RSAs*) (government, administration, army, police, courts, prisons) and the *Ideological State Apparatuses* (*ISAs*) (religion, education, family,

law, politics, trade unions, communication, culture). While the *ISAs* operate via ideology, the *RSAs* operate primarily by force and control. This can be by making illegal the forces and organizations (and their tactics) that threaten the capitalist status quo and the rate of profit. Thus, for example, restrictions are placed on strike action and trade union activities. More extreme versions of *RSA* action include heavy intimidatory policing and other forms of state-sanctioned political repression and violence by the police and armed forces. The ruling class, and governments in whose interests they act, tend to prefer, in normal circumstances, to operate via *ISAs*. Changing the school curriculum to make it more in line with the requirements of capital, for example, is less messy than sending in the riot police or the troops; and it is deemed to be more legitimate by the populace (Hill, 2001b, p. 106; see also Hill, 1989, 2004, 2005b). For Althusser, while the religious *ISA* (system of different churches) used historically to be the major *ISA*, 'the ISA which has been installed in the dominant position in mature capitalist social formations…is the *educational ideological apparatus*…[it is] number one' (Althusser, 1971, p. 153). To what extent, Althusser might have modified his views on education as being the dominant *ISA*, given the current hegemony of organized religion (distortions of Christianity and Islam) is open to debate. In addition, the proliferation of the mass media in its numerous guises might have encouraged Althusser to attribute a more central role to 'culture'. However, there is no denying the current power of education in distilling dominant ideologies. In the United Kingdom, for example, 'education, education, education' was, as is well known, Tony Blair's New Labour mantra. Although there is disagreement among Marxists about the precise relationship between education, privatization, and capitalist profits (e.g., Rikowski, 2005a; Hatcher, 2006; see also Cole, 2007a), there is a general consensus that the education system is a very powerful force in the maintenance of capitalism.

4 Multicultural Education and Antiracist Education in the United States and the United Kingdom

1. Atkinson identifies specifically with *postmodernism;* Baxter identifies with *feminist post-structuralism* and Lather describes herself both as a 'postmodern materialist feminist' (e.g., 1991, p. xix) and as a *feminist post-structuralist* (e.g., 2001). This is not of concern here, since I am interested in their common claim that post-structuralism and postmodernism can be forces for social change and social justice. For a discussion of the origins and theoretical distinctions between post-structuralism and postmodernism, see Cole, 2008d, Chapters 4 and 5.

2. According to Baxter (2002), post-structuralism may ultimately promote social and educational transformation because it listens to all voices, and because it deconstructs. However, as she acknowledges, it is unconvincing in practice and can only (possibly) become convincing if more feminist researchers take it up. No indications are given of how the promotion of transformation might occur. For a further defense of post-structuralism against Marxism, see Baxter (forthcoming, 2009). For my reply, see Cole, 2009c.

3. In so doing, she makes the claim that CRT does 'not necessarily [privilege] race over class, gender, or other identity category' (Ladson-Billings, 2005, p. 57). In chapter 2 of this volume, I have argued that, in fact, such privileging is a defining feature of CRT. Having said that, as recorded above

(endnote 4 of the introduction to this volume), there are a number of off-shoots from CRT which deal with various identities, though not, *crucially*, social class. As we shall see below, the revolutionary multiculturalism of McLaren and others, while centering on class, also encompasses other forms of oppression and exploitation. Later in this chapter, I will note how this has always been the case with antiracist education in the United Kingdom.

4. For a critique of the notion of 'fixed ability', see Hart et al. 2004. For a Marxist assessment of Hart et al., see Cole, 2008b; see also Yarker, 2008; Cole, 2008f.

5. For a recent critical Marxist analysis of *Schooling in Capitalist America*, see Rikowski (2005b).

6. Ladson-Billings, though a Critical Race Theorist, draws at times on post-structuralism and postmodernism and also on Marxism. As I argued in chapter 3, her writing often shows sympathy for Marxist theory and concepts.

7. A turning point was a conference he and I both attended in Halle, in the former DDR in 1995. McLaren has written of our time in Halle:

    My time spent with you was a profound step in getting my work back on track to Marx. No question about that...listening to your talk in Halle, spending time with you, experiencing East Germany with you, that really shook me...it was a major moment, perhaps the key single moment; then, of course, the next step was reading over the criticisms of my work by you, Dave [Hill] and Glenn [Rikowski]...and then the correspondence among us by e-mail...and I began to reeducate myself with the help of comrades such as yourself, reading *Capital*, and a half dozen other books of Marx, Marx and Engels, and also Hegel (via the Marxist humanist tradition of *News and Letters*/Dunayevskaya) putting myself on a program of study and dialogue with other Marxists, working with Ramin [Farahmandpur] and other Marxist students...and I began reading the Open Marxist folks via Glenn, getting into Glenn's work on the labor theory of value...and reading major Marxist journals like *Science and Society*, and then finally meeting Peter Hudis of *News and Letters* and hanging out with him in LA and learning a lot. (personal correspondence, 2002, cited in Cole, 2005. Cole, 2005 contains an analysis of McLaren's trajectory from postmodernism to Marxism: see pp. 108–114)

8. Fryer cites evidence of a large percentage of skeletons of black Africans found among 350 excavated in 1951, dating back to Roman times (Fryer, 1984, cited in Brandt, 1986, p. 7).

9. For an analysis of racism and education from the days of the British Empire up to the present, see, for example Cole and Blair, 2006.

10. In 1979, the nomenclature 'black' in the United Kingdom was generally a term that applied to all people of color and indeed, as a political statement, also used by some South (Mediterranean) Europeans such as Cypriots.

11. The National Front was the United Kingdom's biggest fascist political party at the time, with as many as 20,000 members in the mid-1970s. It declined after 1979 with the advent of the Thatcher Government and the subsequent lurch to the right in British politics.

12. Antiracist education in the United Kingdom has traditionally shown awareness of equality issues other than 'race'. Over twenty years ago, in my own writing for example, (e.g., Cole, 1986a; see also Troyna, 1987), issues of social class and gender/patriarchy were central to the analysis.

13. It should be recorded that there was one mainstream exception. The left-wing Inner London Education Authority (ILEA), eventually abolished by Thatcher in 1988, published a number of equality documents in the 1980s, including *Race, Sex and Class: 4. Anti-Racist Statement and Guidelines* (ILEA, 1983), and distributed the pamphlets to all of its schools. For an analysis of the political climate in the years of the Radical Right, see, for example Hill, 1989, 1997; see also Jones, 2003; Tomlinson, 2005.

14. This was all part of a concerted Radical Right attack on Teacher Education which was assumed to be a hotbed of Marxism (see Hill, 1989, 1994, 2001a, 2001b, 2007a; see also Cole, 2004c, pp. 150–163). This legacy continues to this day. For example, the term 'trainee' rather than 'student teacher' relates to the Radical Right notion that teacher *training* is not a theoretical enterprise, but a combination of love of subject and practical skills. For similar reasons it is the *Training and Development Agency* that oversees teacher education. However, progressive equalities legislation (see Nixon, 2008) has required departments of education in universities, university colleges and colleges to verse their student teachers in equality issues (for detailed suggestions as to how the current QTS standards can be used to have equality and equal opportunities at their forefront, see Cole (2008e); for suggestions on promoting equality in the primary/elementary school, see Hill and Helavaara Robertson (2009), and for ideas for the secondary/high school, see Cole, 2009b.

15. 'PC' or 'Political correctness' is a pernicious concept invented by the Radical Right, which, to my dismay, has become common currency in the United Kingdom. The term was coined to imply that there exist (Left) political demagogues who seek to impose their views on equality issues, in particular appropriate terminology, on the majority. In reality, nomenclature changes over time. Thus, in the twenty-first century, terms such as 'negress' or 'negro' or 'coloured'/'colored', nomenclatures that at one time were considered quite acceptable, and are now considered offensive. Egalitarians are concerned with *respect* for others and, therefore, are careful to acknowledge changes in nomenclature, changes that are decided by oppressed groups themselves, bearing in mind that there can be differences among such oppressed groups. Thus, e.g., it has become common practice to use 'working class' rather than 'lower class'; 'lesbian, gay, bisexual and transgender' rather than 'sexually deviant'; 'disability' rather than 'handicap', 'gender equality' rather than 'a woman's place'. Using current and acceptable nomenclature is about the fostering of a caring and inclusive society, not about 'political correctness' (Cole, 2008d, pp. 142–143).

5   CRT Comes to the United Kingdom: A Critical Analysis of David Gillborn's *Racism and Education*

1. An exception might be 'new multicultural education' or 'radical multicultural education' as advocated, for example, by Mal Leicester (e.g., Leicester, 1992a). Leicester challenged antiracism on a number of grounds from a '[L]eft liberal' perspective, arguing that multicultural and antiracist education could be synthesized (Leicester, 1992b, p. 265). At the time I argued (e.g., Cole, 1992a) that such a synthesis, as advocated by Leicester, was not possible, primarily because her approach seemed to rely on white teachers teaching 'new multicultural' education, and would therefore be most likely to perpetuate stereotypes (e.g., Cole, 1992a, p. 247; for detailed discussions on these

differing approaches, including an interchange of views, see Leicester, 1992a; Cole, 1992a; Leicester, 1992b; Cole, 1992b). I now believe, given modern ICT facilities, that antiracist multicultural education is possible (see chapter 8 of this volume).

2. CRT seems geographically limited. While Ladson-Billings (2006, p. xii) states that '[i]n its adolescence CRT...takes on an international dimension', the only examples she gives, apart from the United States, are the United Kingdom and the suburbs of Paris. Gillborn (2008, p. 1) has argued that, in addition to the United States and United Kingdom, CRT has relevance to Canada, Europe and Australasia. Marxism, on the other hand, is unequivocally international.

3. Notable exceptions with respect to a belief in the need for a wide-ranging definition of racism are, of course, Robert Miles and the Glasgow University sociologists (see chapter 2 of this volume).

4. In some senses the Chronicle resembles Michael Winterbottom's 2005 film-within-a-film, *A Cock and Bull Story*, where Steve Coogan plays Tristram Shandy and Steve Coogan intermittently.

5. Alpesh Maisuria and Chris Martin concluded a recent paper with the following statement:

> We feel an integration of CRT and Marxism is some way off, although they both have something to offer each other. For example, the concept of 'voice' that is so prominent in CRT is something that the Marxian method can develop. In the light of this, while the legitimacy of Marxian critique is not dependent on insights coming from CRT, we believe that a strategic alliance between the two would be mutually beneficial and that both would benefit from healthy debate...In sum, theorists engaged with CRT and Marxism are both striving for social justice, in the light of this, optimism and positive discussion is needed. (Maisuria and Martin, 2008)

6. Allen's (2007) views on 'whiteness' include the following: 'whites do not possess [love of the oppressed and] we whites...are more likely to have disdain or pity, certainly not love, for people of color' (p. 57); '[t]he best a white person can be is a white anti-racist racist' (p. 62).

7. This is not to deny that Marx held racist views or made racist comments. Diane Paul (1981) has examined a number of Marx's (and Engel's) public and private comments on race from a historian's point of view. Paul Warmington (personal correspondence, 2008) has summarized some worthwhile points from Paul's paper:

- Marx's casual racism, particularly in private correspondence, was precisely that: dull casual racism—in the sense that it was unexceptional in a nineteenth-century European context, in which reified bio-cultural understandings of 'race' and belief in racial hierarchies were orthodox, 'everyday' beliefs. Thus Marx's racism cannot be denied but it can hardly be viewed as an integral element of a thought out 'Marxist' political position (this may be less true of his anti-Semitic comments). For example, Marx wrote in casually racist terms in private, but also not only publicly opposed the south in the U.S. Civil War but, at times, took a public pro-Irish standpoint.

- we should not try to transform Marx into twentieth/twenty-first century progressivism on every issue.

- *most importantly*, the key issue for Marxists is not to deny or to agonize over Marx's nineteenth-century racism but to recognize that Marxism is a 'living philosophy', continually being adapted and adapting itself 'by means of thousands of new efforts' (Sartre, 1960).

One of the key founders of CRT, W.E.B. Du Bois was also a product of his time. Du Bois wrote a letter to his friend Aptheker on February 27, 1953 expressing regret for his anti-Semitic references to the Jews ('The jew is the heir of the slave baron in Dougherty'; 'thrifty and avericious Yankee, shrewed and unscrupulous Jews'; 'the enterprising Russian Jew who sold it to him'; 'out of which only a yankee or a Jew could squeeze more blood from debt cursed tenants'):

> And I did not, when writing, realize that by stressing the name of the group instead of what some members of the [group] may have done, I was unjustly maligning a people in exactly the same way my folk were then and are now falsely accused. (Cited in Hayes Edwards, 2007, p. xxvi)

8. I am using 'ahistorical' in its true sense of 'lacking historical perspective or context' (Pearsall, 2001), not in the sense of not referring to history. Gillborn does make references to historical incidents, but does not contextualize them with respect to the prevailing mode of production and accompanying ideology at the time.

9. Of course, some workers are less alienated than others. Gillborn and I, for example, are able to promote our fundamental (albeit theoretically different) beliefs in social justice during the course of our wage labor.

10. This was first published as Bell (1990) and subsequently reprinted a number of times, including a chapter in Bell (1992b).

11. The *Empire Windrush* arrived at Tilbury, a town on the north bank of the River Thames, on June 22, 1948, carrying 492 passengers from Jamaica, the first large group of African-Caribbean migrants to arrive in the United Kingdom after World War II.

12. I have included continent-wide and global institutions in my definition. The former, for example, would incorporate xeno-racism and xeno-racialization in the case of Europe, in the light of the enlargement of the European Union (see chapter 2 of this volume for a discussion). With respect to the global dimension, modes of production throughout the world have become globally racialized in new ways and have created new institutionally racist structures. In an analysis, which is essentially postmodern but informed by Marxism, Bhattacharyya et al. (2002) have provided a number of insights, which can aid in the development of our understanding of a Marxist concept of racialization. Their argument is that, in the global economy, racialization has taken on new forms, for example, the way in which the World Trade Organization has supported multinationals against developing world farmers (p. 30); how 'downsizing' has spawned a revival of old-fashioned sites and forms of production like 'sweatshops' and homeworking (p. 31) and the general way in which high street goods on sale in the West are produced by excessively exploited labor in developing countries (pp. 32–33). Bhattacharyya et al. make interesting observations on new forms of institutional racism, such as the 'prison-industrial complex', which 'performs the oldest management tricks in the book—undercuts wages, exploits the most constrained workforce imaginable and, through this, disrupts the organization of workers on the outside' (2002, p. 43), and on the role of the WTO and the IMF in the debt crisis (pp. 111–136). The spatial proximity of the racialized poor and the rich, they point out, is needed 'because insurance companies, law firms, banks, etc., need cleaners, porters and gardeners: they also need entertainment both for their own workforce and for their international clients' (p. 131). In addition,

they attempt to extend Marxist analysis by noting the way in which bio-technology and genetic modification extends the idea of ownership to the organic world (pp. 116–121) (see also chapter 6 of this volume).

6 Neoliberal Global Capitalism and Imperialism in the Twenty-first Century

1. This chapter draws on and develops parts of Cole, 2008d, Chapters 7 and 8.
2. A blatant example of enfraudening enantiomorphism within the public sphere occurred at the end of 'Britain's Best: 2008', an award ceremony sponsored by the *Sun*, and broadcast on ITV1 on May 23, 2008. Attempts were made to enfrauden the viewers at the climax of the program that featured Gordon Brown and 'The British Armed Forces' (the latter being the winner of the *Sun* global recognition award). We are told that the last twelve months has seen the largest number of British troops on active duty overseas since World War II. Why are these working-class men and women overseas? They are there to 'save the lives of others'. Over footage featuring troops interacting happily with local children, we are told that they are also there to 'provide security, stability and hope to those whose lives have been devastated by the effects of war'. In case there was any doubt that they were peacemakers, the accompanying music was chosen carefully. It was Paul Weller's *You Do Something to Me*. Here are two of the lines of the song: 'You do something to me—somewhere deep inside. I'm hoping to get close to—a peace I cannot find'. Crude propaganda of this kind, is, of course, likely to intensify Islamophobia.
3. I am aware that many postmodernists and post-structuralists would totally reject the following arguments. Perhaps it is in the nature of postmodernism that widely differing views can be contained within one school of thought.
4. It was Sigmund Freud (1914) [1991] who postulated an early stage of primal narcissism. During this time an infant is preoccupied with itself and with its own pleasure, while being oblivious of the needs of others.

7 Marxism and Twenty-first-Century Socialism

1. Although, for all progressive people, the British Welfare State was a major achievement (indeed an economic and political consensus was forged which would last until Prime Minister James Callaghan's 1976 Ruskin College speech, and the onset of the Thatcher government of 1979), the changes were not untainted by the legacy of British Imperialism. As the Report put it, 'with its present rate of reproduction the British race cannot continue, means of reversing the recent course of the birth rate must be found' (paragraph 413). Conflating racism with sexism, the Report assigned to women the role of baby-machines in the service of capitalism and British hegemony. Women were told: '[i]n the next thirty years housewives as Mothers have vital work to do in ensuring the adequate continuance of the British Race and British Ideals in the world' (paragraph 117). A clear example of Beveridge's own racism can be seen in his essay 'Children's allowances and the Race'. In it he stated

> Pride of race is a reality for the British as for other peoples...as in Britain today we look back with pride and gratitude to our ancestors, look back as a nation or as individuals two hundred years and more to the generations

illuminated by Marlborough or Cromwell or Drake, are we not bound also to look forward, to plan society now so that there may be no lack of men or women of the quality of those early days, of the best of our breed, two hundred and three hundred years hence? (cited in Cohen, 1985, pp. 88–89)

2. By way of example, it is worth mentioning the special role of the monarchy in the United Kingdom. In addition to the Political/Government RSA/ISA (Cole, 2008g, 2008h), the monarchy continues to make a major contribution to the ideological role of the state, in that it 'normalises' a massively hierarchical society. As Althusser (1971, p. 138) has argued, above the ensemble of state apparatuses is 'the head of State, the government and the administration'. Althusser (p. 154) has also pointed out that 'the English bourgeoisie was able to 'compromise' with the aristocracy and 'share' State power and the use of the State apparatus with it for a long time'. This 'arrangement' continues, albeit in modified form, up to the present day. In the United Kingdom, the head of State is the Queen. The Monarchy, though not as popular as it used to be—'Princess Di as the fairy princess'—is still thriving. The royal family continue to receive mass attention in the media, much of it favorable. Their outrageous activities are normalized and condoned. For example, the vast wealth they have; their private planes and boats; the large number of servants they employ are all generally portrayed as 'necessary' for such important people, who are seen as an 'asset' to the nation. If one of the princes spends thousands of pounds on a night out—when Prince William split up with Kate Middleton in April 2007, he racked up a £11,000 bar bill in one night—this is reported neutrally by the tabloids, as 'understandable'. The aristocracy and the monarchy are not under any imminent threat of extinction. The Queen receives £7.9 million per year from taxpayers via the Civil List, while the income due to the heir to the throne, The Prince of Wales's Office from the Duchy of Cornwall (created in 1337 by Edward III) amounts to £14.067 million (The Duchy of Cornwall, 2006). His Royal Highness, Prince Phillip, The Duke of Edinburgh, Earl of Merionith, Baron Greenwich *KG* (Knight of the Order of the Garter), KT (Knight of the Order of the Thistle), OM (Order of Merit), GBE (Knight Grand Cross of the Order of the British Empire), *AC*, QSO (Queen's Service Order), *PC* (Prince Consort), well known for his racist 'jokes', is, at the time of writing (May 2008) the subject of a prime-time ITV two-part series, in which black newsreader Sir Trevor McDonald asks a number of obsequious and fawning questions to the duke, whom he calls 'sir'. ITV also put out a program in May in which The Duchess of York dropped in to try to 'persuade a chain smoking, overweight family of six from a Hull council estate [one of the poorest in the country], to adopt a healthier lifestyle' (ITV Web site: http://www.itv.com/CatchUp/Video/default.html?ViewType=5&Filter=19725) The commercial break featured a Government 'advert' showing a poor couple's old car being crushed as a warning to the working class to pay their road tax. On the same night BBC1 had a program detailing the government's plans to get a million people off sick benefit. All the above needs to be seen in the light of Gordon Brown's attempt to control the current crisis in capitalism by keeping public sector workers' pay deals below 2% (boardroom pay at Britain's top companies soared by 37% last year). After ten years of New Labour Government in Britain, the wealthiest 1% of the population owns 21% of the nation's wealth, while the bottom 50% owns 7%; health inequalities have grown in the last decade; poverty continues to blight many children's lives; and 67% of ethnic minority communities

live in the 88 most deprived wards in the country (Compass Direction for the Democratic Left, 2007).

3. The ideological role of The Sun is discussed in various places in this volume. While, as noted in chapter 2 of this volume, it is the most popular working class right-wing tabloid, the Daily Mail is the most popular middle-class, right-wing tabloid (I am using sociological rather than Marxist definitions of social class here: as I have stressed throughout this volume, the former being based on occupation and status; the latter on relationship to the mode of production).

4. This section of the chapter on 'Common Objections to Marxism and a Marxist Response' develops Cole, 2008d, pp. 129–138.

5. Marx's views on religion are well known. As he famously put it: 'Religion is the sigh of the oppressed creature, the heart of a heartless world, and the soul of soulless conditions. It is the opium of the people' (Marx, 1843–1844). The editors of the Marx Internet Archives (MIA) (cited with this extract from Marx) explain that in n the nineteenth century, opium was widely used for medical purposes as a painkiller, and thus Marx's dictum did not connote a delusionary state of consciousness. Although Marx and Marxism have traditionally been associated with atheism, my own view is that this is not necessary. While religion, as opposed to theism (belief in a God or Gods that intervene in the world) or deism (belief in a God who does not intervene in the world) has often been and continues to be form of oppression and conservatism, there are large numbers of people who identify with a religious or spiritual belief who also identify with Marxism or socialism (millions of Roman Catholics in Venezuela for example; see later in this chapter). There are also, of course, many who are atheists or agnostics. Although I am not aligned to any religion, I find no contradiction in my own belief in the existence of some power for good. Whatever our beliefs or lack of beliefs, it is my view that our energies should be devoted to the creation of equality and happiness on earth.

6. Several anecdotes attest to this. One will suffice. I once got talking to two Cuban medical doctors celebrating their return to Havana from a number of months working in Africa. They were clearly very glad to be home. When I asked how they felt about working on 'local wages' in Africa, their joint and immediate answer was unequivocal: 'we're doctors, that's what we do'.

7. For a discussion of utopian socialism, see Cole, 2008d, Chapter 2.

8. Here Marx and Engels are referring to members of communist parties. The words 'communist' and 'communism' are greatly misunderstood. 'Communism' was used by Marx to refer to the stage after socialism when the state would have withered away and when we would live communally. In the period after the Russian Revolution up to the demise of the Soviet Union, the Soviet Union and other Eastern European countries were routinely referred to as '[C]ommunist' in the West. The Soviet Union, following Marx, actually referred to itself as 'socialist'. Socialists critical of the Soviet Union (e.g., Cliff, 1974) have referred to the Soviet Union and other Eastern European countries as 'state capitalist'. In this chapter, and on numerous occasions in my writing, I have distanced myself from 'state capitalism' or Stalinism, and have advocated democratic socialism.

9. Here, we have an ironic twist: the capitalist class and their representatives who used to deride Marxists for what they saw as the metaphysic of 'Marxist economic determinism', for what they (wrongly) perceived was a belief in the inevitability

of social revolution, are the ones who now champion the *inevitability* of global neoliberalism, the accompanying 'world-wide market revolution' and the consequent *inevitability* of 'economic restructuring' (McMurtry, 2000).

10. At the same time, globalization, in reality in existence since the beginnings of capitalism, is hailed as a new and unchallengeable phenomenon, and its omnipresence used ideologically to further fuel arguments about capitalism's inevitability. In ex-UK prime minister Tony Blair's (2005) words, challenging globalization is tantamount to denying that autumn follows summer (King, 2005).

11. Chávez was elected president in 1998, and assumed office in February, 1999. He was re-elected in 2000 and in 2006. There was a failed right-wing coup against him in 2002 (for an analysis, see the video by Bartley and O'Briain, 2003). It is not clear how Critical Race Theorists would view the Bolivarian Revolution. For Marxists, it needs to be contextualized historically with respect to (neoliberal) capitalism and imperialisms (see Ali, 2008, pp. 53–90 for an extended discussion of U.S. maneuverings in pre-Chávez Venezuela).

12. While that which follows is very supportive of Chávez, it needs to be pointed out that there are those on the Left who have less favorable (e.g. Azul, 2008) or more mixed (e.g. Henshall, 2008) views. Thus Steve Henshall (2008, p. 6), for example, writing in *Socialist Worker,* the weekly newspaper of the UK-based Socialist Workers Party, argues that while 'numerous progressive schemes have been set up to improve the lives of the poor...[their] lives...could be considerably better'. Henshall (ibid.) also states that the aforementioned nationalization of SIDOR occurred after three months of workers' struggle, and after the then Minister of Labor, Riviero, had tried to make a deal with the bosses behind the workers' backs (ibid.). According to Henshall (ibid.) eventually the National Guard were sent in to try to end a workers' occupation. 'Solidarity from othe factories quickly spread', Henshall concludes (ibid.), 'and Chávez had to intervene, promising to nationalise the plant while quickly firing Riviero'.

13. Chávez quotes Fidel Castro on this fact: '[d]on't say that! Every time you say that, Bush has you in his sights' (cited in Campbell, 2008, p. 58).

14. As noted above, I had the privilege to teach a course at UBV for a week in 2006. The course was entitled *Introduction To World Systems: Global Imperial Capitalism or International Socialist Equality: Issues, and Implications for Education.* Standards at UBV are very high—with seminar discussions and debate comparing more than favorably with universities in which I have taught in the United Kingdom and around the world. However, at UBV, advanced theory is very much linked to practice—that is, to improving the lives of people in the communities from where the students come. Students are almost 100% working class at UBV. While teaching there, I met a police officer who was studying for his second degree. He told me how the Chávez Government was humanizing the police force. He reckoned that Chávez has the support of about 75% of the Caracas police.

15. One thing that symbolizes the revolution for me was the way in which, at the start of my last seminar at UBV, one of the caretakers arrived to unlock the lecture theatre, and then sat down, listened to and actively contributed to my seminar. His question was what percentage of the British working class did I think were revolutionary socialists. When I told him that the percentage was very very small indeed, he seemed somewhat bemused.

16. One example will give a flavor of current Venezuelan health care. Isabel children's hospital has been open for a year. The head surgeon has said that it is 'more than we could have dreamed of'. He points out that children used to wait up to twelve years for surgery, but since the hospital opened, 1,200 operations have been performed for congenital heart disease. Everything is free, including transport to the hospital. Mothers can stay with their children, and a small hotel is provided for visiting fathers (Campbell, 2008, p. 61).

17. In a letter to *The Guardian* (December 17, 2007), John Green argues:

   Whatever one thinks of Chávez, the vituperative knee-jerk reaction of virtually all the media has been unprecedented. When one remembers all the unsavoury dictatorships that ruled in South America not too long ago, with hardly a word of condemnation from the so-called democratic west, the visceral hatred of Chávez can only be explained as attempted brain-washing by those who fear his challenge to their hegemony. He may not fit our northern idea of the perfect diplomat, but he is attempting to change decades of domination by the US and an arrogant elite who have always seen Miami rather than Caracas as their real home. No one accuses him of setting up offshore bank accounts for himself, of living in palaces or rigging elections, so what's his crime? Perhaps the fact that he is offering hope to those who have been disenfranchised, poor and uneducated for generations and thus challenging global capitalism.

18. The fallacy of 'white Marxism' is underlined by Marxist people of color in South America, Africa and Asia in their millions, both historically and in the present.

8 CRT and Marxism: Some Suggestions for Classroom Practice

1. In concentrating on curricular issues, I am, of course, fully aware that many of the important educational struggles against neoliberal global capitalism and imperialism occur outside the school and the classroom.

2. It needs to be reiterated here that, as noted in chapter 2, Preston argues that CRT needs to be considered alongside Marxism. The following pedagogies, however, are firmly embedded in CRT.

3. The training referred to, though not described, by Vaught and Castagno (2008) (discussed in chapter 2 of this volume) would appear to have some similarities with Racism Awareness Training. A more viable alternative might be to have in-depth conceptual discussions on the meaning of racism (see my suggested definition in chapter 2 of this volume), and to follow this up with personal and practical implications (see Cole, 2008a for a discussion of this suggestion with respect to racism and other equality issues).

4. One of the 2006 presidential election slogans (Chávez won with a landslide) was 'Contra El Diablo [Bush], Contra El Imperialismo [Against the Devil, Against Imperialism]'.

Conclusion

1. Martin Luther King's support for organized labor at this time was not new. A few examples will suffice: in a speech in 1961, King declared, '[o]ur needs are identical with labor's needs...[t]hat is why Negroes support labor's demands and fight laws which curb labor'; in 1962, he supported the coalition of 'the Negro and the forces of labor, because their fortunes are so

closely intertwined' (American Federation of State, County, and Municipal Employees [AFSCME], 2008); in 1965, he stated that '[w]hen...in the thirties the wave of union organization crested over our nation, it carried to secure shores not only itself but the whole society'; in 1967, that '[t]he Negroes pressed into [service occupations] need union protection, and the union movement needs their membership'; also in 1967, he argued that '[i]t is natural for Negroes to turn to the Labor movement'; and finally to the Sanitation workers on strike in Memphis in 1968, he insisted: '[y]ou are demanding that this city will respect the dignity of labor...whenever you are engaged in work that serves humanity and is for the building of humanity, it has dignity and it has worth' (Harrity, 2004).

2. On December 8, 1999, a Memphis jury awarded Coretta Scott King and her family $100 in damages for the conspiracy to murder her late husband (Sheppard, 2006, p. 7). According to the jury, Dr. Martin Luther King Jr. was assassinated by a conspiracy that included agencies of the U.S. government (p. 1). As Sheppard concludes, '[f]rom the beginning it has been clear that the FBI was involved to one degree or another' (p. 7).

3. Sheppard, a contemporary of both King and Malcolm X, makes a similar claim about the changing politics of the latter:

> At the time of their assassinations, both Martin Luther King and Malcolm X were embarking on a course in opposition to the capitalist system. It is clear from reading and listening to their final speeches that they had both evolved to similar conclusions as to capitalism's role in the maintenance of racism. That is why they were neutralized.

Sheppard, who was present at Malcolm X's assassination, argues that there is 'irrefutable proof that the government had the motive to assassinate Malcolm X and the ability, through its...spy operations, to orchestrate his assassination. It is now time to open up all the files of the CIA and the FBI, as well as the thousands of pages of files of the New York City Police Department, so that the truth about the assassination of Malcolm X can be exposed' (Sheppard, 2006).

References

Abbass, T. (2007) 'British South Asians and Pathways into Selective Schooling: Social Class, Culture and Ethnicity'. *British Education Research Journal*, 33 (1), pp. 75–90.

Adams, M. and Burke, P.J. (2006) 'Recollections of September 11 in Three English Villages: Identifications and Self-Narrations' *Journal of Ethnic and Migration Studies*, 32 (6), pp. 983–1003.

Ali, T. (2008) *Pirates of the Caribbean: Axis of Hope*, London: Verso.

Allen, R.L. (2006) 'The Race Problem in the Critical Pedagogy Community', in C.A. Rossatto, R.L. Allen, and M. Pruyn (eds.) *Reinventing Critical Pedagogy: Widening the Circle of Anti-oppressive Education*, Lanham, MD: Rowman and Littlefield.

—— (2007) 'Whiteness and Critical Pedagogy', in Z. Leonardo (ed.) *Critical Pedagogy and Race*, Oxford: Blackwell.

Althusser, L. (1971) 'Ideology and Ideological State Apparatuses' in *Lenin and Philosophy and Other Essays*, London: New Left Books. Available HTTP: http://www.marx2mao.com/Other/LPOE70ii.html#s5 (accessed May 30, 2008).

American Federation of State, County and Municipal Employees (AFSCME) (2008) 'Dr. Martin Luther King, Jr. on Labor', AFSCME. Available HTTP: http://www.afscme.org/about/1550.cfm (accessed May 6, 2008).

Ansley, F.L. (1997) 'White Supremacy (and What We Should Do about It)' in R. Delgado and J. Stefancic (eds.) *Critical White Studies: Looking behind the Mirror*, Philadelphia, PA: Temple University Press.

Apple, M.W. (2005). 'Audit Cultures, Commodification, and Class and Race Strategies in Education'. *Policy Futures in Education*, 3 (4), pp. 379–399.

—— (2006) *Educating the "Right" Way*. New York: Routledge.

Atkinson, E. (2001) 'A Response to Mike Cole's "Educational Postmodernism, Social Justice and Social Change: an incompatible ménage-à-trois"'. *The School Field* 12 (1/2), pp. 87–94.

—— (2002) 'The Responsible Anarchist: Postmodernism and Social Change'. *British Journal of Sociology of Education*, 23 (1), pp. 73–87.

Azul, R. (2008) 'Latin American Presidents Meet with Bolivia on Brink of Civil War'. World Socialist Web Site (WSWS), September 16. Available HTTP: http://www.wsws.org:80/articles/2008/sep2008/boli-s16.shtml (accessed September 17, 2008).

Ball, S. (1990) *Politics and Policymaking in Education*. London: Routledge.

BAMN (Coalition to Defend Affirmative Action, Integration, and Immigrant Rights And Fight for Equality By Any Means Necessary) (2001) 'Asian Pacific

Americans and Affirmative Action: From the Trial Transcript of Professor Frank Wu'. Available HTTP: http://www.bamn.com/doc/2001/010212-apas-and-aa-wu-excerpts.asp (accessed August 9, 2008).

Bartley, K. and D. O'Briain (2003). *The revolution will not be televised* [Video]. Available HTTP: http://video.google.com/videoplay?docid=5832390545689805144 (accessed July 14, 2008).

Bartolovich, C. (2002) 'Introduction', in C. Bartolovich and N. Lazarus (eds.) *Marxism, Modernity and Postcolonial Studies*, Cambridge: Cambridge University Press.

Baxter, J. (2002) 'A Juggling Act: A Feminist Post-structuralist Analysis of Girls' and Boys' Talk in the Secondary Classroom'. *Gender and Education*, 14 (1), pp. 5–19.

——— (forthcoming, 2009) 'A Case for Post-structuralism' in SYMPOSIUM *Marxism and Educational Theory: Origins and Issues* (Mike Cole) discussed by Judith Baxter, Samuel Fassbinder, and Ravi Kumar, with a response by Mike Cole, *Policy Futures in Education*.

BBC News (2004) 'Chaos and Violence at Abu Ghraib'. Available HTTP: http://news.bbc.co.uk/2/hi/americas/3690097.stm (accessed October 20, 2008).

——— (2005a) 'Polish Student Assaulted by Gang'. Available HTTP: http://news.bbc.co.uk/1/hi/england/tees/4710011.stm (accessed July 24, 2006).

——— (2005b) 'Three Suffer Unprovoked Beating'. Available HTTP: http://news.bbc.co.uk/1/hi/england/dorset/4713593.stm (accessed July 24, 2006).

——— (2007) 'Poles in Redditch Hit by Racism'. Available HTTP: http://www.bbc.co.uk/herefordandworcester/content/articles/2007/03/22/breakfast_polish_racism_feature.shtml (accessed January 18, 2008).

——— (2008a) 'The £7-per-hour Jobs Locals Don't Want'. Available HTTP: http://news.bbc.co.uk/1/hi/magazine/7288430.stm (accessed July 27, 2008).

——— (2008b) 'Attack on House Treated as Racist' Available HTTP: http://news.bbc.co.uk/1/hi/northern_ireland/foyle_and_west/7452233.stm (accessed July 27, 2008).

——— (2008c) 'Cash Raised after Racist Attack'. Available HTTP: http://news.bbc.co.uk/1/hi/england/shropshire/7304198.stm (accessed July 27, 2008).

——— (2008d) 'Man in Unprovoked Racist Assault'. Available HTTP: http://news.bbc.co.uk/1/hi/scotland/edinburgh_and_east/7350036.stm (accessed July 27, 2008).

——— (2008e) 'Pole Subjected to "Racist Attack"'. Available HTTP: http://news.bbc.co.uk/1/hi/scotland/north_east/7358197.stm (accessed July 27, 2008).

——— (2008f) 'Prison for Racist Polish Attack'. Available HTTP: http://news.bbc.co.uk/1/hi/scotland/edinburgh_and_east/7433720.stm (accessed July 27, 2008).

——— (2008g) 'Rise in Racist Attacks Reported'. Available HTTP: http://news.bbc.co.uk/1/hi/northern_ireland/7266249.stm (accessed July 27, 2008).

Begg, M. and V. Brittain (2006) *Enemy Combatant: A British Muslim's Journey to Guantanamo and Back*, London: Free Press.

Belfast Today (2006) 'Polish Man Hurt in Vicious Attack'. Available HTTP: http://www.belfasttoday.net/ViewArticle2.aspx?SectionID=3425&ArticleID=1532979 (accessed July 24, 2006).

Bell, D. (1973). *Race, Racism and American Law*, Boston: Little, Brown and Company.

——— (1980) '*Brown v. Board of Education* and the Interest-Convergence Principle'. *Harvard Law Review*, 93, pp. 518–533.

——— (1990) 'After We're Gone: Prudent Speculations on America in a Post-racial Epoch'. *St. Louis University Law Journal*, 34, pp. 393–405.

—— (1992a) *And We Are Not Saved: The Elusive Quest for Racial Justice,* New York: Basic Books.

—— (1992b) *Faces at the bottom of the Well: The Permanence of Racism,* New York: Basic Books.

—— (2004) *Silent Covenants: Brown v. Board of Education and the Unfulfilled Hopes for Racial Reform,* New York: Oxford University Press.

Benton, T. (ed.) (1996) *The Greening of Marxism,* New York: Guilford Press.

BERA (2007) *BERA Annual Conference: Final Programme,* Institute of Education, University of London, September 5–8, Edinburgh, UK: BERA Conference Organizers.

Beratan, G. (forthcoming, 2008) 'The Song Remains the Same: Transposition and the Disproportionate Representation of Minority Students in Special Education'. *Race, Ethnicity and Education,* 11(4), pp. 337–354.

Beveridge, W. (1942) *Social Insurance and Allied Services (The Beveridge Report)* London: HMSO.

Beyer, L.E. and D.P. Liston (1992) 'Discourse or Moral Action? A Critique of Postmodernism'. *Educational Theory,* 42 (4), pp. 371–393.

Bhattacharyya, A. (2008) 'The Movement behind Obama'. *Socialist Worker,* November 15. Available HTTP: http://www.socialistworker.co.uk/art.php?id= 16399 (accessed November 19, 2008).

Bhattacharyya, G., J. Gabriel and S. Small (2002) *Race and Power: Global Racism in the Twenty-first Century,* London: Routledge.

Blair, M. (1994) 'Black Teachers, Black Pupils/Students and Education Markets'. *Cambridge Journal of Education,* 24, pp. 277–291.

Brayboy, B. (2005) 'Toward a Tribal Critical Race Theory in Education'. *The Urban Review,* 37 (5) December, pp. 425–446.

Boulangé, A. (2004) 'The Hijab, Racism and the State'. *International Socialism* (102), Spring, pp. 3–26.

Bowles, S. and H. Gintis (1976) *Schooling in Capitalist America,* London: Routledge and Keegan Paul.

Brandt, G. (1986) *The Realization of Anti-racist Teaching,* Lewes, UK: Falmer Press.

British Venezuela Solidarity Campaign (2006). *Bolivarian Achievements: Social Missiones.* Available HTTP: http://www.venezuelasolidarity.org.uk/ven/web/ 2006/missiones/social_missiones.html (accessed March 31, 2008).

Bukharin, N. (1922) 'Theorie des Historischen Materialismus'. *The Communist International,* pp. 343–345.

Callinicos, A. (1989) *Against Postmodernism: A Marxist Critique,* Cambridge: Polity Press.

—— (1993) *Race and Class,* London: Bookmarks.

—— (2000) *Equality,* Oxford: Polity Press.

Camara, B. (2002) 'Ideologies of Race and Racism', in P. Zarembka (ed.) *Confronting 9-11, Ideologies of Race, and Eminent Economists,* Oxford: Elsevier Science.

Campbell, N. (2008) 'When the Supermodel Met the Potentate'. *GQ Magazine,* February, pp. 56–61.

Campbell, D. and S. Goldenberg (2004); Afghan Detainees Routinely Tortured and Humiliated by US Troops'. *The Guardian,* June 23. Available HTTP: http://www. guardian.co.uk/world/2004/jun/23/usa.afghanistan3 (accessed July 14, 2008).

Carby, H.V. (1979) *Multicultural Fictions,* Stencilled Occasional Paper SP No. 58, University of Birmingham: Centre for Contemporary Cultural Studies.

Castles, S. and G. Kosack (1985) *Immigrant Workers and Class Structure in Western Europe*, Oxford: Oxford University Press.

Centre for Contemporary Cultural Studies (CCCS) (1977) *On Ideology*, London: Hutchinson.

———— (1982) *The Empire Strikes Back: Race and Racism in 70s Britain*, London: Hutchinson.

Centre for Contemporary Cultural Studies (CCCS) Education Group (1981) *Unpopular Education: Schooling and Social Democracy in England since 1944*, London: Hutchinson.

Centre for Contemporary Cultural Studies (CCCS) Women's Studies Group (1978) *Women Take Issue: Aspects of Women's Subordination*, London: Hutchinson.

Chakrabarty, N. (2006a) 'Cultural Mendacity and the Educational Uncanny'. Paper presented at day seminar, Critical Race Theory, Manchester Metropolitan University (November 20, 2006) and at Discourse, Power Resistance 6, at Manchester Metropolitan University, Manchester (March 26, 2007).

———— (2006b) 'On Not Becoming White: The Impossibility of Multiculturalism While Race Is Fixed/Fucked'. Clandestino, Gothenburg, Sweden (Performance), June 8, 2006.

Chakrabarty, N. and J. Preston (2006) 'Posturing Fear in a World of Performed Evil: Terrorists, Teachers, and Evil Neo-liberals'. Conference paper presented at 7th Global Conference Perspectives on Evil and Human Wickedness Salzburg, Austria (March 16, 2006), and 8th Annual Comparative Literature Conference: Cultures of Evil and the Attractions of Villainy. University of South Carolina (February 10, 2006).

———— (2007) 'Putnam's Pink Umbrella? Transgressing Melancholic Social Capital', in H. Helve and J. Bynner (eds.) *Youth and Social Capital*, London: Tufnell Press.

Chang, R.S. (2000) 'Toward an Asian American Legal Scholarship: Critical Race Theory, Post-structuralism, and Narrative Space, in R. Delgado and J. Stefancic (eds.) *Critical Race Theory: The Cutting Edge* (2nd Edition), Philadelphia, PA: Temple University Press.

Chang, R.S. and N. Gotanda (2007) 'Afterword: The Race Question in LatCrit Theory and Asian American Jurisprudence'. *Nevada Law Journal* 7, November 16, pp. 1012–1029. Available HTTP: http://nevadalawjournal.org/pdf/NVJ314.pdf (accessed August 9, 2008).

Chouhan, K. (2006) 'Letter from the Black London Forum and the 1990 Trust'. Available HTTP: http://www.blink.org.uk/docs/CEHR_priorities.pdf (accessed January 22, 2007).

Clarke, J., C. Critcher, and R. Johnson (eds.) (1979) *Working Class Culture: Studies in History and Theory*, London: Hutchinson.

Cliff, T. (1974) *State Capitalism in Russia*. Available at HTTP: http://www.marxists.org/archive/cliff/works/1955/statecap/index.htm (accessed July 6, 2008).

Coben, D. (2002) 'Metaphors for an Educative Politics: "Common Sense," "Good Sense" and Educating Adults', in C. Borg, J. Buttigieg, and P. Mayo (eds.) *Gramsci and Education*, Lanham, MD: Rowman and Littlefield.

Cohen, S. (1985) 'Anti-Semitism, Immigration Controls and the Welfare State', *Critical Social Policy*, 13, Summer.

Cole, M. (1986a) 'Multicultural Education and the Politics of Racism in Britain'. *Multicultural Teaching*, Autumn.

—— (1986b) 'Teaching and Learning about Racism: A Critique of Multicultural Education in Britain', in S. Modgil, G.K. Verma, K. Mallick, and C. Modgil (eds.) *Multicultural Education: The Interminable Debate,* Barcombe, UK: Falmer Press.

—— (1989) 'Monocultural, Multicultural and Antiracist Education' in M. Cole (ed.) *The Social Contexts of Schooling,* Barcombe, UK: Falmer Press.

—— (1992a) 'British Values, Liberal Values or Values of Justice and Equality? Three Approaches to Education in a Multicultural Society', in J. Lynch, C. Modgil, and S. Modgil (eds.) *Equity or Excellence?: Education and Cultural Reproduction Cultural Reproduction and the Schools: Vol. 3,* London: Falmer Press.

—— (1992b) 'Cole Replies to Leicester', in J. Lynch, C. Modgil, and S. Modgil (eds.) *Equity or Excellence?: Education and Cultural Reproduction Cultural Reproduction and the Schools: Vol. 3,* London: Falmer Press.

—— (1992c) *Racism, History and Educational Policy: From the Origins of the Welfare State to the Rise of the Radical Right.* Unpublished Ph.D. thesis, University of Essex.

—— (1998a) 'Globalisation, Modernisation and Competitiveness: A Critique of the New Labour Project in Education'. *International Studies in the Sociology of Education,* 8 (3), 315–332.

—— (1998b) 'Racism, Reconstructed Multiculturalism and Antiracist Education'. *Cambridge Journal of Education,* 28 (1), Spring, pp. 37–48.

—— (1998c) 'Re-establishing Antiracist Education: A Response to Short and Carrington', *Cambridge Journal of Education,* 28 (2), Summer, pp. 235–238.

—— (2004a) ' "Brutal and Stinking" and "Difficult to Handle": The Historical and Contemporary Manifestations of Racialisation, Institutional Racism, and Schooling in Britain'. *Race, Ethnicity and Education,* 7 (1), pp. 35–56.

—— (2004b) 'F*** You—Human Sewage: Contemporary Global Capitalism and the Xeno-racialization of Asylum Seekers'. *Contemporary Politics,* 10 (2), pp. 159–165.

—— (2004c) 'Rethinking the Future: The Commodification of Knowledge and the Grammar of Resistance' in M. Benn and C. Chitty (eds.) *For Caroline Benn: Essays in Education and Democracy,* London: Continuum.

—— (2004d) ' "Rule Britannia" and the New American Empire: A Marxist Analysis of the Teaching of Imperialism, Actual and Potential, in the British School Curriculum'. *Policy Futures in Education,* 2 (3 and 4), pp. 523–538. Available HTTP: http://www.wwwords.co.uk/pfie/content/pdfs/2/issue2_3.asp#7 (accessed July 28, 2008).

—— (2005) 'The "Inevitability of Globalized Capital" vs. the "Ordeal of the Undecidable": a Marxist critique', in M. Pruyn and L.M. Heurta-Charles (eds.) *Teaching Peter McLaren: Paths of Dissent,* New York: Peter Lang.

—— (2006a) 'The Eclipse of the Non-European: Historical and Current Manifestations of Racialization in the US', *Ethnic and Racial Studies,* 29 (1) 2006, pp. 173–185.

—— (2006b) ' "Looters and Thugs and Inert Women Doing Nothing": Racialized Communities in Capitalist America and the Role of Higher Education'. *Journal for Critical Education Policy Studies,* 4 (1), March. Available at: http://www.jceps.com/index.php?pageID=article&articleID=64 (accessed August 4, 2007).

—— (2007a) 'Neo-liberalism and Education: A Marxist Critique of New Labour's Five Year Strategy for Education', in A. Green, G. Rikowski, and H. Raduntz (eds.) *Renewing Dialogues in Marxism and Education: Volume 1—Openings*, Basingstoke, UK: Palgrave Macmillan.

Cole, M. (2007b) ' "Racism" Is about More than Colour'. *Times Higher Education (THE)* November 23.

——— (2007c) 'Reality Racism: Challenging Racism in the Era of Big Brother'. *EnglishDramaMedia*, October, pp. 9–15 and pp. 35–38.

——— (2008a) 'Introductory Chapter: Education and Equality—Conceptual and Practical Considerations', in M. Cole (ed.) *Professional Attributes and Practice: Meeting the QTS Standards* (4th edition), London: Routledge.

——— (2008b) '*Learning without Limits*: A Marxist Assessment'. *Policy Futures in Education*, 6 (4), pp. 453–463.

——— (2008c) 'Maintaining "the Adequate Continuance of the British Race and British Ideals in the World": Contemporary Racism and the Challenges for Education', Inaugural Professorial Lecture, delivered November 13, 2006, Bishop Grosseteste University College Lincoln.

——— (2008d) *Marxism and Educational Theory: Origins and Issues*, London: Routledge.

——— (ed.) (2008f) *Professional Attributes and Practice for Student Teachers: Meeting the QTS Standards* (4th edition), London: Routledge.

——— (2008g) 'Reply to Pat Yarker'. *Policy Futures in Education*, 6 (4), pp. 468–469.

——— (2008h) 'The State Apparatuses and the Working Class: Experiences from the UK; Educational Lessons from Venezuela', in D. Hill (ed.) *Contesting Neoliberal Education: Public Resistance and Collective Advance*, New York: Routledge.

——— (2008i) 'The Working Class and the State Apparatuses in the UK and Venezuela: Implications for Education'. *Educational Futures* 2.

——— (2009a) 'Critical Race Theory Comes to the UK: A Marxist Critique'. *Ethnicities*.

——— (ed.) (2009b) *Equality in the Secondary School: Promoting Good Practice across the Curriculum*, London: Continuum.

——— (2009c) 'On Capitalism, Imperialism, Social Justice and Socialism', in SYMPOSIUM *Marxism and Educational Theory: Origins and Issues* (Mike Cole) discussed by Judith Baxter, Samuel Fassbinder, and Ravi Kumar, with a response by Mike Cole, *Policy Futures in Education*. Forthcoming 2009.

——— (2009d) 'A Response to Charles Mills'. *Ethnicities*.

Cole, M. and M. Blair (2006) 'Racism and Education: From Empire to New Labour', in M. Cole (ed.) *Education, Equality and Human Rights: Issues of Gender, 'Race', Sexuality, Disability and Social Class* (2nd edition), London: Routledge.

Cole, M. and A. Maisuria (2007) ' "Shut the F*** Up", "You Have No Rights Here": Critical Race Theory and Racialisation in Post-7/7 Racist Britain'. *Journal for Critical Education Policy Studies* 5 (1).

——— (2008) 'Racism in Post 7/7 Britain: Critical Race Theory, Marxism and Empire: What Is the Role of Education?', in D. Kelsh, D. Hill, and S. Macrine (eds.) *Class in Education: Knowledge, Pedagogy, Subjectivity*, New York: Routledge.

Cole, M. and S. Virdee (2006) 'Racism and Resistance: From Empire to New Labour', in M. Cole (Ed.) *Education, Equality and Human Rights: Issues of Gender, 'Race', Sexuality, Disability and Social Class* (2nd edition), London: Routledge.

Commission on Integration and Cohesion (2007) Our Shared Future, Wetherby, UK: Communities and Local Government.

Compass Direction for the Democratic Left (2007) *Closer to Equality? Assessing New Labour's Record on Equality after 10 Years in Government,* London: Compass, Direction for the Democratic Left.

Contreras Baspineiro, A. (2003) 'Globalizing the Bolivarian Revolution Hugo Chávez's Proposal for Our América'. Available HTTP: http://www.narconews. com/Issue29/article746.html (accessed July 25, 2007).

Cooper, R. (2002) 'The Post Modern State'. Available HTTP: http://www.fpc.org. uk/articles/169 (accessed July 6, 2008).

Corriente Marxista Revolucionaria (Venezuela) (2008) 'President Chávez on National TV Twice Quotes Alan Woods' Book on Bolshevism', *In Defence of Marxism,* June 2. Available HTTP: http://www.marxist.com/chavez-quotes-al-an-woods-book-bolshevism.htm (accessed June 3, 2008).

Crenshaw, K. (1988) 'Race, Reform and Retrenchment: Transformation and Legitimation in Antidiscrimination Law'. *Harvard Law Review,* 10, pp. 1331–1387.

—— (1995) 'Mapping the Margins; Intersectionality, Identity Politics, and Violence against Women of Color', in K. Crenshaw, N. Gotanda, G. Peller, and K. Thomas (eds.) *Critical Race Theory: The Key Writings that Formed the Movement,* New York: New Press.

Crenshaw, K., N. Gotanda, G. Peller, and T. Kendall, T. (1995a) 'Critical Race Theory and Critical Legal Studies: Contestation and Coalition', in K. Crenshaw, N. Gotanda, G. Peller, and K. Thomas (eds.) *Critical Race Theory: The Key Writings that Formed the Movement,* New York: New Press.

—— (1995b) 'Introduction', in K. Crenshaw, N. Gotanda, G. Peller, and K. Thomas (eds.) *Critical Race Theory: The Key Writings that Formed the Movement,* New York: New Press.

Cruddas, J., P. John, N. Lowles, H. Margetts, D. Rowlands, D. Shutt, and S. Weir (2005) 'The Far Right in London: A Challenge for Local Democracy?: York'. Joseph Rowntree Reform Trust. Available HTTP: http://www.jrrt.org.uk/Far_Right_REPORT.pdf (accessed August 5, 2008).

Cruz, C.Z. (2005) 'Four Questions on Critical Race Praxis: Lessons from Two Young Lives in Indian Country'. *Fordham Law Review,* 73 (5), pp. 2133–2160.

Curran, F. (2007) 'What Happened to the Global Justice Movement? Fighting Neoliberalism—The View from Scotland'. *International Viewpoint.* Available HTTP: http://www.internationalviewpoint.org/spip.php?article1365 (accessed July 29, 2008).

Dale, G. (1999) 'Capitalism and Migrant Labour', in G. Dale and M. Cole (eds.) *The European Union and Migrant,* Oxford: Berg.

Dale, G. and M. Cole (eds.) (1999) *The European Union and Migrant Labour,* Oxford: Berg.

Dallmayr, F. (2004) 'The Underside of Modernity: Adorno, Heidegger, and Dussel'. *Constellations,* 11 (1), pp. 102–120.

Dalton, H.L. [1987] (1995) 'The Clouded Prism: Minority Critique of the Critical Legal Studies Movement', in K. Crenshaw, N. Gotanda, G. Peller, and K. Thomas (eds.) *Critical Race Theory: The Key Writings that Formed the Movement,* New York: New Press.

Daniel, W.W. (1968) *Racial Discrimination in England,* Harmondsworth, UK: Penguin.

Darder, A. and R.D. Torres (2004) *After Race: Racism after Multiculturalism,* New York and London: New York University Press.

Davidson, N. (2006) 'The 1926 General Strike: Nine Days of Hope'. *Socialist Worker* Available HTTP: http://www.socialistworker.co.uk/art.php?id=8752 (accessed May 10, 2008).

Davis, D. (2008) 'Myths of Intelligence'. in *Socialist Worker,* May 31.

Davis, R. (2008) ' "You Live in a Caravan? Wow!" How Can Schools Make Traveller Children Feel at Home?' *The Guardian,* May 27. Available HTTP: http://education.guardian.co.uk/egweekly/story/0,,2282220,00.html (accessed October 20, 2008).

DeCuir, J.T. and A.D. Dixson (2004) ' "So When It Comes Out, They Aren't that Surprised that It Is There": Using Critical Race Theory as a Tool of Analysis of Race and Racism in Education'. *Educational Researcher,* 33 (5), June/July, pp. 26–31.

DeCuir v. Benson (1875) 27 La. Ann. 1 (Fifth District Court of Orleans Parish).

DeCuir-Gunby, J.T. (2006) ' "Proving Your Skin Is White, You Can Have Everything": Race, Racial Identity, and Property Rights in Whiteness in the Supreme Court Case of Josephine DeCuir', in A.D. Dixson and C.K. Rousseau (eds.) *Critical Race Theory in Education: All God's Children Got a Song,* New York and London: Routledge.

Dehal, I. (2006) Powerpoint available HTTP: http://www.blss.portsmouth.sch.uk/training/ppt/DFES_ID_KD.ppt#13 (accessed January 2, 2009).

Delgado, R. (1984) [1995] 'The Imperial Scholar: Reflections on a Review of Civil Rights Literature', in K. Crenshaw, N. Gotanda, G. Peller, and K. Thomas (eds.) *Critical Race Theory: The Key Writings that Formed the Movement,* New York: New Press.

——— (1992) 'The Imperial Scholar Revisited: How to Marginalize Outsider Writing, Ten Years Later'. *University of Pennsylvania Law Review,* 140, pp. 1349–1372.

——— (1995) *The Rodrigo Chronicles: Conversations about America and Race,* New York: New York University Press.

——— (1998) 'Rodrigo's Committee Assignment: A Sceptical Look at Judicial Independence'. *Southern California Law Review,* 72, pp. 425–454.

——— (2001) 'Two Ways to Think about Race: Reflections on the Id, the Ego, and Other Reformist Theories of Equal Protection'. *Georgetown Law Review* 89. Available HTTP: http://findarticles.com/p/articles/mi_qa3805/is_200107/ai_n8985367/pg_2 (accessed March 26, 2008).

——— (2003) 'Crossroads and Blind Alleys: A Critical Examination of Recent Writing about Race'. *Texas Law Review,* 121.

Delgado, R. and J. Stefancic (eds.). (2000) *Critical Race Theory: The Cutting Edge* (2nd edition), Philadelphia, PA: Temple University Press.

——— (2001) *Critical Race Theory: An Introduction,* New York: New York University Press.

Demie, F. and R. Tong (2007) *Education Statistics.* London: Lambeth Research and Statistics Unit.

Demie, F., K. Lewis, and C. McLean (2007) *Raising the Achievement of Somali Pupils: Challenges and School Responses.* London: Lambeth Research and Statistics Unit.

Democratic Socialists of Central Ohio (2008) Available HTTP: http://dscol.tripod.com/ (accessed July 29, 2008).

Department for Children, Schools and Families (2005) 'Higher Standards, Better Schools for All'. Available HTTP: http://www.dfes.gov.uk/publications/schoolswhitepaper/ (accessed May 26, 2008).

Department of Children, Schools and Families Standards Site (2008) 'Gypsy, Roma and Traveller Achievement'. Available HTTP: http://www.standards.dfes.gov.uk/ethnicminorities/raising_achievement/gypsy_travellerachievement/ (accessed August 5, 2008).

Derrida, J. (1990) 'Force of Law: The Mystical Foundation of Authority', translated by Mary Quaintance. *Cardozo Law Review*, 11, pp. 919–1070.

de Siqueira, A.C. (2005) 'The Regulation of Education through the WTO/GATS'. *Journal for Critical Education Policy Studies*, 3 (1), March. Available HTTP: www.jceps.com/?pageID=article&articleID=41 (accessed July 6, 2008).

Devidal, P. (2009) 'Trading Away Human Rights? The GATS and the Right to Education: A Legal Perspective', in D. Hill and R. Kumar (eds.) *Global Neoliberalism and Education and its Consequences*. New York: Routledge.

Dixson, A. D. (2006) 'The Fire This Time: Jazz Research and Critical Race Theory', in A. D. Dixson and C. K. Rousseau (eds.) *Critical Race Theory in Education: All God's Children Got a Song, New York and London: Routledge.*

Dixson, A.D. and C.K. Rousseau (2006a) 'And We Are Still Not Saved: Critical Race Theory in Education Ten Years Later', in A.D. Dixson, and C.K. Rousseau (eds.) *Critical Race Theory in Education: All God's Children Got a Song*, New York and London: Routledge.

———— (2006b) 'Introduction', in A.D. Dixson and C.K. Rousseau (eds.) *Critical Race Theory in Education: All God's Children Got a Song*, New York and London: Routledge.

Dodd, V. (2005) 'Asian Men Targeted in Stop and Search. Huge Rise in Number Questioned Under Anti-Terror Laws', *The* Guardian, August 17. Available HTTP: www.guardian.co.uk/attackonlondon/story/0,16132,1550470,00.html (accessed July 28, 2007).

Dresser, M. (1989) 'The Colour Bar in Bristol, 1963', in R. Samuel (ed.) *Patriotism: The Making and Unmaking of British National Identity, Vol. 1 History and Politics*, London: Routledge.

Doughty, L. (2008) 'History Repeating'. *The Guardian*, September 16. Available HTTP: http://www.guardian.co.uk/world/2008/sep/16/roma.race (accessed September 17, 2008).

Du Bois, W.E.B. (1903) 'The Forethought', in *The Souls of Black Folk*. Available HTTP: http://www.bartleby.com/114/100.html (accessed July 24, 2007).

———— (1915) *The Negro*, New York: Henry Holt.

Eagleton, T. (1991) *Ideology*, London: Verso.

Education and Inspections Act (2006), London: HM Government. Available HTTP: http://www.opsi.gov.uk/Acts/acts2006/ukpga_20060040_en_1 (accessed July 28, 2008).

Elliott, L. (2008) 'Saving Fannie and Freddie Was Nationalisation Pure and Simple: It's the Free-marketeers Who Are to Blame but, by Not Seizing the Moment, It's the Left that Could Well End Up Carrying the Can'. *The Guardian*, September 9. Available HTTP: http://www.guardian.co.uk/commentisfree/2008/sep/09/freddiemacandfanniemae.subprimecrisis (accessed September 12, 2008).

Engels, F. (1890) 'The Materialist Conception of History: From a Letter to Joseph Bloch', London, September 21. Available HTTP: www.socialistparty.org.uk/WhatIsMarxFrame.htm (accessed July 6, 2008).

———— (1892) [1977] 'Socialism: Utopian and Scientific', in *Karl Marx and Frederick Engels: Selected Works in One Volume*, London: Lawrence and Wishart.

Engels, F. (1895) [1972] 'Marx-Engels Correspondence 1890 to J. Bloch in Königsberg', in *Historical Materialism (Marx, Engels, Lenin)*, Moscow: Progress Publishers. Available HTTP: http://www.marxists.org/archive/marx/works/ 1890/letters/ 90_09_21.htm (accessed May 20, 2008).

Espinoza, L. and A.P. Harris (2000) 'Embracing the Tar-Baby: LatCrit Theory and the Sticky Mess of Race' in R. Delgado and J. Stefancic (eds.) *Critical Race Theory: The Cutting Edge*, Philadelphia, PA: Temple University Press.

Fanon, F. (1963) *The Wretched of the Earth*, New York: Grove.

Fekete, L. (2009) *A Suitable Enemy: Islamophobia and Xeno-racism in Europe*, London: Pluto Press.

Feldman, P. and C. Lotz (2004) *A World to Win: A Rough Guide to a Future without Global Capitalism*, London: Lupus Books.

Fenton, S. (2003) *Ethnicity*, Cambridge: Polity Press.

Ferguson, N. (2003) Prince and Empire Are the Key to History, *The Sunday Times*, July 6.

Ferguson, N. (2004) 'American Empire—Who Benefits?', *Empire and the Dilemmas of Liberal Imperialism*, CD accompanying *Prospect*, March 2004.

Ferguson, N. (2005) 'Admit It, George Dubya's Medicine Is Not all Bad'. *Times Higher Education Supplement*, March 18, 2005. Available HTTP: http://www. timeshighereducation.co.uk/story.asp?storyCode=194801&sectioncode=26 (accessed July 6, 2008).

Foster, J.B. (2000) *Marx's Ecology: Materialism and Nature*, New York: Monthly Review Press.

—— (2002) *Ecology against Capitalism*, New York: Monthly Review Press.

Freud, S. (1914) [1991] 'On Narcissism: An Introduction' *Contemporary Freud: Turning Points and Critical Issues*, New Haven, CT: Yale University Press.

Fryer, P. (1984) *Staying Power: The History of Black People in Britain*, London: Pluto Press.

—— (1988) *Black People in the British Empire: an Introduction*, London: Pluto Press.

Gair, R. (2006) 'Ellis faces disciplinary charges'. Available HTTP: http://campus. leeds.ac.uk/newsincludes/newsitem3675.htm (accessed July 14, 2008).

Galloway, G. (2006) 'Address to Marxism 2006'. *Forum: What Next for Respect*, CD 51, London: Bookmarks. Available HTTP: www.bookmarks.uk.com/cgi/ store/bookmark.cgi (accessed October 20, 2008).

Gatswatch (undated). Available HTTP: http://www.gatewatch.org/ (accessed November 29, 2008).

General Union of Oil Employees in Basr (2008) 'Mayday Message from Iraqi Trade Unions' General Union of Oil Employees in Basra', May 1. Available HTTP: http://www.basraoilunion.org/ (accessed July 6, 2008).

Getty, S. (2006) 'East Enders Say They Love London'. *The Metro*.

Gibson, R. and G. Rikowski (2004) *Socialism and Education: An E-Dialogue*, conducted between July 19–August 8, at Rich Gibson's *Education Page for a Democratic Society*. Available HTTP: http://www.pipeline.com/~rougeforum/ RikowskiGibsonDialogueFinal.htm (accessed July 18, 2008).

Gibson-Graham, J.K. (1996) 'Querying Globalization'. *Rethinking Marxism*, 9, Spring, pp. 1–27.

Gillborn, D. (2005) 'Education Policy as an Act of White Supremacy: Whiteness, Critical Race Theory and Education Reform', *Journal of Education Policy*, 20 (4), July, pp. 485–505.

———— (2006a) 'Critical Race Theory and Education: Racism and Antiracism in Educational Theory and Praxis'. *Discourse: Studies in the Cultural Politics of Education*, 27 (1), pp. 11–32.

———— (2006b) 'Rethinking White Supremacy: Who Counts in "WhiteWorld"'. *Ethnicities*, 6 (3), pp. 318–340.

———— (2008) *Racism and Education: Coincidence or Conspiracy?* London: Routledge.

Gillborn, D and H. Mirza (2000) *Educational Inequality; Mapping Race, Class and Gender—A Synthesis of Research Evidence,* London: Ofsted.

Gilroy, P. (1987) *There Ain't No Black in the Union Jack,* London: Hutchinson.

———— (2004) *After Empire: Melancholia or Convivial Culture?* Oxfordshire, UK: Routledge.

Glenn, J. and L. Barnett (2007) 'Spate of Racist Complaints against Polish Pupils'. *Times Educational Supplement* March 30.

Goldenberg, S. (2008) US Election 2008: '"I Want to Cut His Nuts Out"—Jackson Gaffe Turns Focus on Obama's Move to the Right'. *The Guardian,* July 11. Available HTTP: http://www.guardian.co.uk/world/2008/jul/11/barackobama.uselections2008 (accessed July 11, 2008).

Gotanda, N. (1995) 'A Critique of "Our Constiution Is Color-Blind"', in K. Crenshaw, N. Gotanda, G. Peller, and K. Thomas (eds.) *Critical Race Theory: The Key Writings that Formed the Movement,* New York: New Press.

Gramsci, A. (1921) 'Unsigned, L'Ordine Nuovo', March 4, 1921, text from Antonio Gramsci *Selections from Political Writings (1921–1926),* translated and edited by Quintin Hoare (London: Lawrence and Wishart, 1978), transcribed to the World Wide Web with the kind permission of Quintin Hoare. Available HTTP: http://www.marxists.org/archive/gramsci/1921/03/officialdom.htm (accessed July 6, 2008).

———— (1978) *Selections from Prison Notebooks,* London: Lawrence and Wishart

Gruenwald, D.A. (2003) 'The Best of Both Worlds: A Critical Pedagogy of Place', *Educational Researcher,* 32, pp. 3–12.

Haberman, M. (1991) 'The Pedagogy of Poverty versus Good Teaching', *Phi Delta Kappan,* 73, pp. 290–294.

Hall, S., C. Critcher, T. Jefferson, J. Clarke, and B. Robert (1978) *Policing the Crisis: Mugging, the State and Law and Order,* London: Macmillan.

Hall, S. (1978) 'Racism and Reaction', in BBC/CRE *Five Views of Multi-Racial Britain,* London: BBC/CRE.

Hall, S. and T. Jefferson (1976) *Resistance Through Rituals: Youth Subcultures in Post-war Britain,* London: Hutchinson.

Hall, S., D. Hobson, A. Lowe, and P. Willis (eds.) (1980) *Culture, Media, Language,* London: Hutchinson.

Hampton, P. (2006) 'Chávez Wins Election, but What about the Workers'. Available HTTP: http://www.workersliberty.org/node/7388 (accessed July 27, 2008).

Hands off Iraqi Oil (2008) 'The Attack on Labour Rights Today: Iraqi, US and UK Union Solidarity against War, Occupation and Enforced Privatisation', June 26. Available HTTP: http://www.handsoffiraqioil.org/search?q=the+attack+on+labour+rights+today/ (accessed November 29, 2008).

Hare, B. (2006) 'Law Symposium to Feature Black Racial Theorists', *The Michigan Daily* Tuesday, January 24, 2006. Available HTTP: http://www.michigandaily.com/media/paper851/news/2005/02/04/News/Law-Symposium.To.

Feature.Black.Racial.Theorists-1428501.shtml?norewrite&sourcedomain=www.
michigandaily.com (accessed July 9, 2008).

Harman, C. (1996) 'Globalization: A Critique of a New Orthodoxy', *International Socialism*, no. 73, pp. 3–33.

―――― (2008) 'Economic Crisis: Capitalism Exposed', *Socialist Review*, 322, February, pp. 10–13.

Harris, C. (1993) 'Whiteness as property', *Harvard Law Review*, 106, pp. 1707–1791.

Harris, L.C. (2007) 'Real Rights and Recognition Replace Racism in Venezuela', 13 July. Available HTTP: http://www.venezuelasolidarity.org/?q=node/372 (accessed July 25, 2007).

Harris, P. (2008) 'Forty Years after the Shot Rang Out, Race Fears Still Haunt the US' http://www.guardian.co.uk/world/2008/mar/30/race.uselections2008 (accessed October 28, 2008).

Harrity, J. (2004) 'Dr. King Dedicated his Life to the Causes of Working People' *International Association of Machinists and Aerospace Workers*, January 16. Available HTTP: http://www.goiam.org/content.cfm?cID=3273 (accessed July 29, 2008).

Hart, S., A. Dixon, M.J. Drummond, and D. McIntyre (2004) *Learning without Limits*, Maidenhead: Open University Press.

Hatcher, R. (2006) 'Privatisation and Sponsorship: the re-agenting of the school system in England', *Journal of Education Policy*, 21 (5), pp. 599–619.

Hayes Edwards, B. (2007) *The Souls of Black Folk—Collection of 14 essays by W.E.B. Du Bois*, Oxford Oxford University Press.

Heath, A. and Ridge, J. (1983) 'Social Mobility of Ethnic Minorities'. *Journal of Biosocial Science*, Supplement 8, pp. 169–184.

Herrera, L.E. (2002) 'Challenging a Tradition of Exclusion: The History of an Unheard Story at Harvard Law School', *Harvard Latino Law Review*, 5, Spring, pp. 51–140.

Herrnstein, R.J. and C. Murray (1994) *The Bell Curve*, New York: Free Press.

Hickey, T. (2002) 'Class and Class Analysis for the Twenty-first Century', in: M. Cole (ed.) *Education, Equality and Human Rights*, London: Routledge/Falmer.

―――― (2006) ' "Multitude" or "Class": Constituencies of Resistance, Sources of Hope', in: M. Cole (ed.) *Education, Equality and Human Rights* (2nd edition), London: Routledge.

Hill, D. (1989) *Charge of the Right Brigade: The Radical Right's attack on Teacher Education*, Brighton: Institute for Education Policy Studies. Available HTTP: http://www.ieps.org.uk.cwc.net/hill1989.pdf (accessed July 5, 2008).

―――― (1994) Initial Teacher Education and Ethnic Diversity. In G. Verma and P. Pumfrey (eds.), *Cultural Diversity and the Curriculum, Vol 4: Cross-Curricular Contexts, Themes and Dimensions in Primary Schools*. London: Falmer Press.

―――― (1997) 'Equality in Primary Schooling: The Policy Context, Intentions and Effects, of the Conservative "Reforms" ', in M. Cole, D. Hill and S. Shan (eds.), *Promoting Equality in Primary Schools*, London: Cassell.

―――― (2001a) 'Equality, Ideology and Educational Policy', in D. Hill and M. Cole (eds.) *Schooling and Equality: Fact, Concept and Policy*, London: Kogan Page.

―――― (2001b) State Theory and the Neo-Liberal Reconstruction of Schooling and Teacher Education: A Structuralist Neo-Marxist Critique of Postmodernist, Quasi-Postmodernist, and Culturalist Neo-Marxist Theory, *The British Journal of Sociology of Education*, 22 (1), pp. 137–157.

―――― (2003) 'Global Neo-Liberalism, the Deformation of Education and Resistance'. *Journal for Critical Education Policy Studies*, 1 (1). Available HTTP:

http://www.jceps.com/index.php?pageID=article&articleID=7 (accessed October 28, 2008).

—— (2004) 'Books, Banks and Bullets: Controlling Our Minds—The Global Project of Imperialistic and Militaristic Neo-liberalism and Its Effect on Education Policy'. *Policy Futures in Education,* 2 (3–4) (Theme: Marxist Futures in Education). Available HTTP: http://www.wwwords.co.uk/pfie/content/pdfs/2/issue2_3.asp (accessed October 28, 2008).

—— (2005a) 'Globalisation and Its Educational Discontents: Neoliberalisation and Its Impacts on Education Workers' Rights, Pay, and Conditions'. *International Studies in the Sociology of Education,* 15 (3), pp. 257–288.

—— (2005b) 'State Theory and the Neoliberal Reconstruction of Schooling and Teacher Education', in G. Fischman, P. McLaren, H. Sünker, and C. Lankshear (eds.) *Critical Theories, Radical Pedagogies and Global Conflicts.* Boulder, CO: Rowman and Littlefield.

—— (2007a) 'Critical Teacher Education, New Labour in Britain, and the Global Project of Neoliberal Capital'. *Policy Futures,* 5 (2), pp. 204–225. Available HTTP: http://www.wwwords.co.uk/pfie/content/pdfs/5/issue5_2.asp (accessed October 28, 2008).

—— (2007b) *'Education, Class and Capital in the Epoch of Neo-liberal Globalisation',* in A. Green, G. Rikowski, and H. Raduntz (eds.) *Marxism and Education: Renewing Dialogues: Volume 1—Opening the Dialogue.* London: Palgrave Macmillan.

—— (2008a) 'A Marxist Critique of Culturalist/Idealist Analyses of "Race", Caste and Class'. *Radical Notes* http://radicalnotes.com/content/view/68/39 (accessed October 28, 2008).

—— (2008b) ' "Race", Class and Neoliberal Capital', in C. Mallot and B. Porfilio (eds.) *An International Examination of Urban Education: The Destructive Path of Neoliberalism,* Rotterdam, The Netherlands: Sense.

—— (ed.) (2009a) *The Rich World and the Impoverishment of Education: Diminishing Democracy, Equity and Workers' Rights.* New York: Routledge.

—— (ed.) (2009b) *Contesting Neoliberal Education: Public Resistance and Collective Advance,* London: New York: Routledge.

—— (2009c, forthcoming) Caste, 'Race' and Class: A Marxist Critique of Caste Analysis, Critical Race Theory; and Equivalence (or Parallellist) Explanations of Social Inequality'. *Cultural Logic.*

Hill, D. and E. Rosskam (eds.) (2009) *The Developing World and State Education: Neoliberal Depredation and Egalitarian Alternatives.* New York: Routledge.

Hill, D. and L. Helavaara Robertson, (eds.) (2009), *Equality in the Primary School: Promoting Good Practices across the Curriculum,* London: Continuum.

Hill, D. and R. Kumar (eds.) (2009) *Global Neoliberalism and Education and Its Consequences.* New York: Routledge.

Hill, D., N. Greaves, and A. Maisuria (2008) 'Education, Inequality and Neoliberal Capitalism: A Classical Marxist Analysis'. in D. Hill and R. Kumar (eds.) (2009) *Global Neoliberalism and Education and Its Consequences.* New York: Routledge.

Hill, D, P. McLaren, M. Cole, and G. Rikowski (eds.) (2002) *Marxism against Postmodernism in Educational Theory,* Lanham, MD: Lexington Books.

Hill, M. (1997) *After Whiteness: Unmasking an American Majority,* New York: NYU Press.

Hillcole Group (1997) *Rethinking Education and Democracy: A Socialist Alternative for the Twenty-first Century,* London: Tufnell Press.

Hiro, D. *Black British, White British,* London: Pelican.

Holmes, C. (1979) *Anti-Semitism in British Society 1876–1939,* London: Edward Arnold.

The Home Office (2000) *Race Relations Amendment Act,* London: HMSO.

hooks, b. (1989) *Talking Back: Thinking Feminist. Thinking Black,* Boston, MA: South End Press.

The Human Rights Act (1998) Available HTTP: http://www.direct.gov.uk/en/RightsAndResponsibilities/Citizensandgovernment/DG_4002951 (accessed March 27, 2008).

Ignatiev, N. and J. Garvey (eds.) (1996) *Race Traitor,* New York: Routledge.

Inner London Education Authority (ILEA) (1983) *Race, Sex and Class: 4. Anti-Racist Statement and Guidelines,* London: ILEA.

Irvin, G. (2008) *Super Rich: The Rise of Inequality in Britain and the United States, Cambridge: Polity Press.*

Isaksen, J.L. (2000) 'From Critical Race Theory to Composition Studies; Pedagogy and Theory Building'. *Legal Studies Forum* 24 (3 and 4), pp. 695–711.

Jones, K. (2003) *Education in Britain 1944 to the Present.* Cambridge: Polity Press.

Jones, K. (2008) 'US-Pakistani Relations Remain on the Boil'. World Socialist Web Site (WSWS) September 20. Available HTTP: http://www.wsws.org:80/articles/2008/sep2008/paki-s20.shtml (accessed October 20, 2008).

Kahn, R. (2003) 'Paulo Freire and Eco-Justice: Updating Pedagogy of the Oppressed for the Age of Ecological Calamity'. Available HTTP: http://getvegan.com/ecofreire.htm (accessed 28 July 2007).

Kay, J. (2008) 'US Military Deaths in Iraq Reach 4,000: Eight US soldiers and Dozens of Iraqis Killed in Weekend Violence'. World Socialist Web Site March 24. Available HTTP: http://www.wsws.org/articles/2008/mar2008/iraq-m24.shtml (accessed March 24, 2008).

Kelsh, D. and D. Hill (2006) 'The Culturalization of Class and the Occluding of Class Consciousness: The Knowledge Industry in/of Education'. *Journal for Critical Education Policy Studies,* 4 (1). Available HTTP: http://www.jceps.com/index.php?pageID=article&articleID=59 (accessed October 22, 2008).

Kelsh, D., D. Hill, and S. Macrine (eds.) (2009) *Teaching Class: Knowledge, Pedagogy, Subjectivity.* London: Routledge.

Kennedy, D. (1982) 'Legal Education and the Reproduction of Hierarchy'. *Legal Education,* 32, pp. 591–615.

Kidder, W.C. (2003) 'The Struggle for Access from *Sweatt* to *Grutter*: A History of African American, Latino, and American Indian Law School Admissions, 1950–2000'. *Harvard BlackLetter Law Journal,* 19, pp. 1–42.

King, J.E. (2001) 'Culture-Centered Knowledge: Black Studies, Curriculum Transformation, and Social Action', in J.A. Banks and C.A.M. Banks (eds.) *Handbook of Research on Multicultural Education,* San Francisco: Jossey Bass.

King, S. (2005) 'It's All Very Well Beating the French, but It's the Chinese we Really Have to Worry About'. *The Independent,* October 3. Available HTTP: http://www.independent.co.uk/news/business/comment/stephen-king-its-all-very-well-beating-the-french-but-its-the-chinese-we-really-have-to-worry-about-509388.html/ (accessed November 28, 2008).

Kinnear, M. and S. Barlow (2005) 'The Eco Crisis and What It Means'. Paper presented at *A Climate for Change—Tackling the Eco Crisis and Corporate Power* Conference, October. Available HTTP: www.aworldtowin.net/about/c4cConf.html (accessed July 9, 2008). The Web site of the organization, *A World to Win,* which organized the Conference is www.aworldtowin.net.

Kirk, N. (1985) *The Growth of Working Class Reformism in Mid-Victorian England*, London: Croom Helm.

Kovel, J. (1988) *White Racism: A Psychohistory*, London: Free Association Books.

Ladson-Billings, G. (2005) 'New Directions in Multicultural Education: Complexities, Boundaries, and Critical Race Theory', in J.A. Banks and C.A.M. Banks (eds.) *Handbook of Research on Multicultural Education*, San Francisco: Jossey Bass.

—— (2006). 'Foreword They're Trying to Wash Us Away: The Adolescence of Critical Race Theory in Education', in A.D. Dixson and C.K. Rosseau (eds.) *Critical Race Theory in Education: All God's Children Got a Song*, New York: Routledge.

Ladson-Billings, G. and W.F. Tate (1995) 'Toward a Critical Race Theory of Education'. *Teachers College Record*, 97 (1), pp. 47–68.

Lantier, A. (2008) 'Obama's Transition: A Who's Who of Imperialist Policy'. *World Socialist* Web Site (WSWS). Available HTTP: http://www.wsws.org/articles/2008/nov2008/pers-n19.shtml (accessed November 19, 2008).

Lasky, J. (2003) 'Marx and Engels on the Civil War', *Crossfire*, 73. Available HTTP: http://www.americancivilwar.org.uk/news_marx-engels-on-the-civil-war_11. htm (accessed June 1, 2008).

Lather, P. (1991) *Getting Smart; Feminist Research and Pedagogy with/in the Post-modern*, New York: Routledge.

—— (2001) 'Ten Years Later, Yet Again: Critical Pedagogy and Its Complicities', in K. Weiler (ed.) *Feminist Engagements: Reading, Resisting and Revisioning Male Theorists in Education and Cultural Studies*, London: Routledge.

Lavalette, M., G. Mooney, E. Mynott, K. Evans, and B. Richardson (2001) 'The Woeful Record of the House of Blair'. *International Socialism*, 90. Available HTTP: http://pubs.socialistreviewindex.org.uk/isj90/lavalette.htm (accessed February 14, 2008).

Lawrence, E. (1982) 'Just Plain Common Sense: The "roots" of Racism', in Centre for Contemporary Cultural Studies (ed.) *The Empire Strikes Back: Race and Racism in 70s Britain*. London: Hutchinson.

Lawrence III, C.R., M.J. Matsuda, R. Delgado, and K. Williams Crenshaw (1993) 'Introduction', in M.J. Matsuda, C.R. Lawrence III, and R. Delgado (eds.) *Words that Wound: Critical Race Theory, Assaultive Speech and the First Amendment*, Boulder, Colorado: Westview Press.

Leicester, M. (1992a) 'Antiracism versus New Multiculturalism: Moving beyond the Interminable Debate', in J. Lynch, C. Modgil, and S. Modgil (eds.) *Cultural Diversity and the Schools. Vol. 3 Equity or Excellence? Education and Cultural Reproduction*, London: Falmer Press.

—— (1992b) 'Leicester Replies to Cole', in J. Lynch, C. Modgil, and S. Modgil (eds.) *Equity or Excellence?: Education and Cultural Reproduction Cultural Reproduction and the Schools: Vol. 3*, London: Falmer Press.

Lee, F.J.T. (2005) 'Venezuela's President Hugo Chavez Frias: "The Path is Socialism"'. Available: HTTP: http://www.handsoffvenezuela.org/chavez_path_socialism_4.htm (accessed May 4, 2007).

Leonardo, Z. (2004) 'The Unhappy Marriage between Marxism and Race Critique: Political Economy and the Production of Racialized Knowledge'. *Policy Futures in Education*, 2 (3 and 4), pp. 483–493. Available HTTP: http://www.wwwords. co.uk/pfie/content/pdfs/2/issue2_3.asp#4 (accessed July 9, 2008).

Lind, M. (2004) Debate: After the War, CD accompanying *Prospect*, March 2004.

Long, W.R. (2005) 'Critical Legal Studies; The Viet Nam Generation Comes of Age'. February 11. Available HTTP: http://www.drbilllong.com/Jurisprudence/CLS.html (accessed August 1, 2008).

Luce, E. (2008) 'Obama under Fire over Iraq Troop Pledge'. *Financial Times*, June 24. Available HTTP: http://www.ft.com/cms/s/0/292c39c0–4217-11dd-a5e8–0000779fd2ac.html (accessed June 30, 2008).

Luxemburg, Rosa. (1916) 'The War and the Workers-The Junius Pamphlet'. Available HTTP: http://h-net.org/~german/gtext/kaiserreich/lux.html (accessed July 9, 2008).

Lynn, M. (2007) 'Critical Race Theory, Afrocentricity, and Their Relationship to Critical Pedagogy', in Z. Leonardo (ed.) *Critical Pedagogy and Race*, Oxford: Blackwell.

Mac an Ghaill, M. (2000) 'The Irish in Britain: The Invisibility of Ethnicity and Anti-Irish Racism'. *Journal of Ethnic and Migration Studies*, 26(1), 137–147.

MacKenzie, J.M. (1988) *Propaganda and Empire: The Manipulation of British Public Opinion 1880–1960*, Manchester: Manchester University Press.

Macpherson, W. (1999) *The Stephen Lawrence Enquiry, Report of an Enquiry by Sir William Macpherson*, London: HMSO. Available HTTP: http://www.archive.official-documents.co.uk/document/cm42/4262/sli-06.htm#6.6 (accessed May 27, 2008).

Maisuria, A. (2006) 'A Brief History of the British "Race" Politics and the Settlement of the Maisuria Family'. *Forum: For Promoting 3–19 Comprehensive Education*, 48 (1), pp. 95–101.

Maisuria, A. and C. Martin (2008) 'Missed Opportunities? The Limitations of Social Justice as Conceptualised in Critical "Race" Theory'. Paper delivered at the *Race-ing Forward* conference, Sunley Management Centre, University of Northampton.

Mandel, E. (1970 [2008]) 'Bourgeois ideology and proletarian class consciousness' in Leninist Organisation—Part 2. Available HTTP: http://www.international viewpoint.org/spip.php?article464 (accessed August 14, 2008).

Marable, M. (2004) 'Globalization and Racialization'. Available HTTP: http://www.zmag.org/content/showarticle.cfm?SectionID=30&ItemID=6034 (accessed July 24, 2007).

Marshall, G. (1998) *A Dictionary of Sociology*, Oxford: Oxford University Press. Available HTTP: http://www.encyclopedia.com/doc/1O88-alienation.html (accessed August 11, 2008).

Martin, J. (2007) 'Chávez Recommends the Study of Trotsky, Praises the Transitional Programme'. *In Defence of Marxism*, April 26. Available HTTP: http://www.marxist.com/chavez-transitional-programme.htm (accessed June 3, 2008).

——— (2008) 'Chávez Renationalises SIDOR—Victory for the Workers!' April 16. Available HTTP: http://www.handsoffvenezuela.org:80/chávez_renationalises_sidor.htm (accessed June 3, 2008).

Martin, P. (2008a) 'As War Clouds Gather: Democrats Back Covert US Attacks on Iran'. World Socialist Web Site (WSWS) June 30. Available HTTP: http://www.wsws.org:80/articles/2008/jun2008/iran-j30.shtml (accessed July 1, 2008).

Martin, P. (2008b) 'Forty Years On, Some Lessons from the Life—and Death—of Dr. Martin Luther King Jr.' World Socialist Web Site (WSWS) April 7. Available HTTP: http://www.wsws.org:80/articles/2008/apr2008/king-a07.shtml (accessed May 6, 2008).

Martinez, E. and A. García (2000) 'What Is "Neo-Liberalism" A Brief Definition'. *Economy 101*. Available HTTP: http://www.globalexchange.org/campaigns/econ101/neoliberalDefined.html (accessed July 11, 2008).

Marx, K. (1843–1844) 'Introduction to a Contribution to the Critique of Hegel's Philosophy of Right'. Available HTTP: http://www.marxists.org/archive/marx/works/1843/critique-hpr/intro.htm (accessed October 22, 2008).

——— (1844) *Economic and Philosophical Manuscripts*. Available HTTP: http://www.marxists.org/archive/marx/works/1844/manuscripts/labour.htm (accessed May 19, 2008).

——— (1845) [1976] 'Theses on Feuerbach', in C.J. Arthur (ed.) *Marx and Engels, The German Ideology*, London: Lawrence and Wishart.

——— (1847) [1995] *The Poverty of Philosophy*, Loughton: Prometheus Books.

——— (1859) [1977] Preface to *A Contribution to the Critique of Political Economy*, Moscow: Progress Publishers. Available HTTP: http://www.marxists.org/archive/marx/works/1859/critique-pol-economy/preface.htm (accessed August 5, 2008).

——— 1860 [1985] 'Marx to Engels, after 11 January 1860', in *Marx and Engels Collected Works*, Vol. 41, London: Lawrence and Wishart.

——— (1862) 'A London Workers' Meeting', *Marx and Engels Collected Works, Vol. 19*. Available HTTP: http://www.marxists.org/archive/marx/works/1862/02/02.htm (accessed June 1, 2008).

——— (1875) [1996] *Critique of the Gotha Programme*, Beijing: Foreign Language Press.

——— (1887) [1965] *Capital, Vol. 1*, Moscow: Progress Publishers. Available HTTP: http://www.maxists.org/archive/marx/works/1847/poverty-philosophy/index.htm/ (accessed December 30, 2008).

Marx, K. and F. Engels (1845) [1975] 'The Holy Family'. Available HTTP: http://www.marxists.org/archive/marx/works/1845/holy-family/ch04.htm (accessed July 27, 2008).

——— (1847) [1977a] 'The Communist Manifesto', in *Karl Marx and Frederick Engels: Selected Works in One Volume*, London: Lawrence and Wishart.

Mason, B. (2008) 'Islamophobia in the British Media'. World Socialist Web Site (WSWS). Available HTTP: http://www.wsws.org:80/articles/2008/jul2008/isla-j28.shtml (accessed July 28, 2008).

Maunder, J. (2006) 'Marxism and the Global South'. *Socialist Worker*, June 17.

McGreal, C. (2008a) 'Mbeki Condemns Violence as a "Disgrace"'. The *Guardian*, May 26. Available HTTP: http://www.guardian.co.uk/world/2008/may/26/southafrica.zimbabwe (accessed June 3, 2008).

——— (2008b) 'Thousands Seek Sanctuary as South Africans Turn on Refugees'. *The Guardian*, May 20. Available HTTP: http://www.guardian.co.uk/world/2008/may/20/zimbabwe.southafrica (accessed May 20, 2008).

McIntosh, P. (1988) 'White Privilege and Male Privilege: A Personal Account of Coming to See Correspondences through Work in Women's Studies'. Working Paper #189, Wellesley College Center for Research on Women, Wellesley, MA 02481.

McLaren, P. (1994) 'White Terror and Oppositional Agency: Towards a Critical Multiculturalism', in D.T. Goldberg (ed.) *Multiculturism: A Critical Reader*, Cambridge, MA: Blackwell.

——— (1995) *Critical Pedagogy and Predatory Culture: Oppositional Politics in a Postmodern Era*, London and New York: Routldege.

——— (1997) *Revolutionary Multiculturalism: Pedagogies of Dissent for the New Millenium*, Boulder, CO: Westview Press.

——— (2000) *Che Guevara, Paulo Freire and the Pedagogy of Revolution*. Oxford: Rowman and Littlefield.

——— (2003) *Life in Schools* (4th edition), Boston MA: Pearson Education.

McLaren, P. (2008) 'This Fist Called My Heart: Public Pedagogy in the Belly of the Beast'. *Antipode* 40(3), pp. 472–481.

McLaren, P. and R. Farahmandpur (2005) *Teaching against Global Capitalism and the New Imperialism: A Critical Pedagogy*, Oxford: Rowman and Littlefield.

McLaren P. and D. Houston (2005) 'Revolutionary Ecologies: Ecosocialism and Critical Pedagogy', in P. McLaren *Capitalists and Conquerors: A Critical Pedagogy Against Empire*, Lanham MD: Rowman and Littlefield.

McMurtry, J. (1998) *Unequal Freedoms: The Global Market as an Ethical System*, Toronto: Garamond Press.

—— (2000) 'Education, Struggle and the Left Today'. *International Journal of Educational Reform*, 10 (2), pp. 145–162.

—— (2002) *Value Wars: The Global Market versus the Life Economy*, London: Pluto Press.

Meiksins Wood, E. (1998) 'Modernity, Postmodernism or Capitalism?' in R.W. McChesney, E., M. Wood, and J.B. Foster (eds.) *Capitalism and the Information Age*, New York: Monthly Review Press.

—— (2003) *Empire of Capital*, London and New York: Verso.

Metro (2007) ' "Pikey" Is Now a Race Hate Word'. Available HTTP: http://www.metro.co.uk/news/article.html?in_article_id=79744&in_page_id=34 (accessed October 22, 208).

Miles, R. (1982) *Racism and Migrant Labour*, London: Routledge and Kegan Paul.

—— (1984) 'Marxism versus the Sociology of Race Relations'. *Ethnic and Racial Studies*, 7 (2), April, pp. 217–237.

—— (1987) *Capitalism and Unfree Labour: Anomaly or Necessity?* London: Tavistock.

—— (1989) *Racism*, London: Routledge.

—— (1993) *Racism after 'Race Relations'*, London: Routledge.

Mills, C. W. (1997) *The Racial Contract*, New York: Cornell University Press.

—— (2003) *From Class to Race: Essays in White Marxism and Black Radicalism*, Lanham, MD: Rowman and Littlefield.

—— (2007) 'Reply to Critics' in C. Pateman and C.W. Mills (eds.) *Contract and Domination*, Cambridge: Polity Press.

—— (2009, forthcoming) 'Critical Race Theory: A Reply to Mike Cole'. *Ethnicities*.

Molyneux, J. (2008) 'Is the World Full Up?' *Socialist Worker*, July 5, p. 13.

Morris, J. (2001) 'Forgotten Voices of Black Educators: Critical Race Perspectives on the Implementation of a Desegregation Plan', *Educational Policy*, 15, pp. 575–600.

Muir, H. (2006a) 'Black Teachers Face Bullying and Racism, Survey Finds'. *The Guardian*, September 8, p. 15.

Murji, K. and J. Solomos (eds.) (2005) *Racialization: Studies in Theory and Practice*, Oxford: Oxford University Press.

Murphy, P. (1995) 'A Mad, Mad, Mad, Mad World Economy'. *Living Marxism*, 80 (June), pp. 17–19.

Nixon, J. (2008) 'Statutory Frameworks Relating to Teachers' Responsibilities', in M. Cole (ed.) *Professional Attributes and Practice for Student Teachers; Meeting the QTS Standards* (4th edition), London: Routledge.

North, D. (2003) 'The Crisis of American Capitalism and the War against Iraq'. *World* Socialist Web Site, March 21. Available HTTP: http://www.wsws.org/articles/2003/mar2003/iraq-m21.shtml (accessed March 24, 2008).

Oborne, P. and Jones, J. (2008) *Muslims under Siege*, Cambridge: Democratic Audit.

Parenti, M. (1998) *America Besieged*, San Francisco: City Lights Books.

Pateman, C. (1988) *The Sexual Contract,* Stanford: Stanford University Press.

Pateman, C. and Mills, C. W. (2007) 'Contract and Social Change', in C. Pateman and C.W. Mills (eds.) *Contract and Domination,* Cambridge: Polity Press.

Pearsall, J. (ed.) (2001) *Concise Oxford Dictionary,* Oxford: Oxford University Press.

Perea, J. (1997) 'Panel: Latina/o Identity and Pan-Ethnicity: Toward Lat Crit Subjectivities: Five Axioms in Search of Equality'. *Harvard Latino Law Review,* 231, Fall, 1997, pp. 231–237.

Perea, J.F., R. Delgado, A.P. Harris, J. Stefancic, and S.M. Wildman (2007) 'Race and Races, Cases and Resources for a Diverse America'. Santa Clara Univ. Legal Studies Research Paper No. 07–25.

Pilkington, A. (2007) 'From Institutional Racism to Community Cohesion: The Changing Nature of Racial Discourse in Britain'. Inaugural professorial lecture, University of Northampton, November 29.

Piper, S. (2007a) 'After the Elections: A New Party for the Venezuelan Revolution'. *International Viewpoint,* January. Available HTTP: http://www.internationalviewpoint.org/spip.php?article1188&var_recherche=chavez (accessed May 17, 2007).

—— (2007b) 'Venezuela: The Challenge of Socialism in the 21st century'. *Socialist Outlook,* 12. Available HTTP: http://www.internationalviewpoint.org/spip.php?article1269 (accessed October 22, 2008).

Platt, L. (2007) *Poverty and Ethnicity in the UK.* Available HTTP: http://www.jrf.org.uk/bookshop/eBooks/2006-ethnicity-poverty-UK.pdf (accessed July 27, 2008).

Postone, M. (1996) *Time, Labor and Social Domination: A Reinterpretation of Marx's Critical Theory,* Cambridge: Cambridge University Press.

Poulantzas, N. (1978) *State, Power, Socialism,* London: Verso.

Preston, J. (2007) *Whiteness and Class in Education,* Dordrecht, The Netherlands: Springer.

Prince, R. and G. Jones (2004) 'My Hell in Camp X-Ray'. Available HTTP: http://www.commondreams.org/headlines04/0507-05.htm (accessed July 19, 2007).

Proyect, L. (2007) 'Gypsy Caravan'. Available HTTP: http://louisproyect.wordpress.com/2007/05/03/gypsy-caravan/ (accessed May 15, 2008).

Puxon, G. (2005) Ustiben Report *Uk Anti-Gypsy Racism Reaches Danger Level.* Available HTTP: http://onevodrom.blogspot.com/2005_03_01_archive.html (accessed October 22, 2008).

QAA (2007a) *New Opportunities: Citizenship Key Stage 3.* Available HTTP: http://www.qca.org.uk/secondarycurriculumreview/subject/ks3/citizenship/planning/opportunities/index.htm (accessed May 6, 2007).

—— (2007b) *Personal, Learning and Thinking Skills in Citizenship Key Stage 3.* Available HTTP: http://www.qca.org.uk/secondarycurriculumreview/subject/ks3/citizenship/links/personal-learning/citizenship/index.htm (accessed May 6, 2007).

—— (2007c) *Programme of Study: Citizenship Key Stage 4.* Available HTTP: http://www.qca.org.uk/secondarycurriculumreview/subject/ks4/citizenship/index.htm (accessed May 6, 2007).

Race Traitor (2005) 16, Winter. Available HTTP: http://racetraitor.org/: (accessed July 27, 2008).

Ramdin, R. (1987) *The Making of the Black Working Class in Britain,* London: Gower.

redhotcurry.com (2005) 'It's Racism, but Not as We Know It', July 21. Available HTTP: http://www.redhotcurry.com/archive/news/2005/cre_annual_report. htm (accessed July 2, 2008).

Reed, A. (2005a) 'The Real Divide' *The Progressive*, November 2005, pp. 27–32. Available HTTP: http://blythe-systems.com/pipermail/nytr/Week-of-Mon-20051107/026287.html (accessed August 11, 2008).

Reed, A. (2005b) 'Class-ifying the Hurricane.' *The Nation*, 281 (10), October 3, pp. 6, 8.

Renton, D. (2006) 'Does Capitalism Need Racism?' Paper presented at the *Racism and Marxist Theory* Workshop, University of Glasgow, September 7–8.

Respect the Unity Coalition (undated) *Another World Is Possible*. Available HTTP: http://www.respectcoalition.org/pdf/f473.pdf (accessed July 10, 2008).

Rifkin, J. (1998) *The Biotech Century*, New York: Tarcher Putnam.

Rikowski, G. (1997) 'Scorched Earth: Prelude to Rebuilding Marxist Educational Theory'. *British Journal of Sociology of Education*, 18 (4), pp. 551–574.

——— (2001) 'The Importance of Being a Radical Educator in Capitalism Today'. Guest Lecture in Sociology of Education, The Gillian Rose Room, Department of Sociology, University of Warwick, Coventry, May 24. Institute for Education Policy Studies. Available HTTP: http://www.ieps.org.uk.cwc.net/rikowski2005a. pdf (accessed August 1, 2008).

Rikowski, G. (2004) 'Marx and the Education of the Future'. *Policy Futures in Education* 2 (3 and 4), pp. 559–571. Available HTTP: http://www.wwwords. co.uk/pdf/viewpdf.asp?j=pfie&vol=2&issue=3&year=2004&article=10_ Rikowski_PFEO_2_3-4_web&id=195.93.21.133 (accessed July 18, 2008).

——— (2005a) 'Distillation: Education in Karl Marx's Social Universe'. Lunchtime Seminar, School of Education, University of East London, Barking Campus, February 14. Available HTTP: http://www.flowideas.co.uk/?page=articles& sub=Distillation (accessed October 22, 2008).

——— (2005b) *Silence on the Wolves: What Is Absent in New Labour's Five Year Strategy for Education*, Brighton, UK: University of Brighton, Education Research Centre, Occasional Paper, May (available from ERC, University of Brighton, Mayfield House, Falmer, Brighton, BN1 9PH; e-mail: Education.Research@brighton.ac.uk).

Roediger, D. (2006) 'The Retreat from Race and Class'. *Monthly Review*, July-August. Available HTTP: http://www.monthlyreview.org/0706roediger.htm (accessed August 11, 2008).

Rose, S. and H. Rose (2005) 'Why We Should Give Up on Race: As Geneticists and Biologists Know, the Term No Longer Has Meaning'. *The Guardian*, April 9. Available HTTP: http://www.guardian.co.uk/comment/story/0,,1455685,00. html (accessed July 27, 2008).

Rousseau, C.K. (2006) 'Keeping It Real; Race and Education in Memphis', in A.D. Dixson and C.K. Rousseau (eds.) *Critical Race Theory in Education: All God's Children Got a Song*, New York and London: Routledge.

Rutter, J. (2006) *Refugee Children in the UK*, Buckingham, UK: Open University Press.

Salomon, A. (1935) 'Max Weber's Political Ideas' *Social Research*, 2, pp. 368–384.

San Juan Jr., E. (2004) 'Post-9/11 Reflections on Multiculturalism and Racism'. *Axis of Logic*, November 13. http://www.axisoflogic.com/cgi-bin/exec/view. pl?archive=79&num=13554.

———. (2008) 'From Race/Racism To Class Struggle: On Critical Race Theory' The Philippines Matrix Project: Interventions toward a National-Democratic

Socialist Transformation. Available HTTP: http://philcsc.wordpress.com/ 2008/10/04/from-raceracism-to-class-struggle-on-critical-race-theory/ (accessed November 21, 2008).

Sands, P. (2008) 'Stress, Hooding, Noise, Nudity, Dogs'. *The Guardian,* April 19. Available HTTP: http://www.guardian.co.uk/world/2008/apr/19/humanrights. interrogationtechniques (accessed May 9, 2008).

Sartre, J.P. (1960) *The Search for Method (1st part). Introduction to Critique of Dialectical Reason.* Available HTTP: http://www.marxists.org/reference/archive/ sartre/works/critic/sartre1.htm (accessed August 8, 2006).

Sarup, M. (1986) *The Politics of Multicultural Education,* London: Routledge and Kegan Paul.

———— (1988) *An Introductory Guide to Post-Structuralism and Postmodernism,* London: Harvester Wheatsheaf.

Scatamburlo-D'Annibale, V. and P. McLaren (2004) 'Class Dismissed? Historical Materialism and the Politics of "Difference"', *Educational Philosophy and Theory,* 36(2), pp. 183–199.

———— (2008) 'Contesting the New "Young Hegelians": Interrogating Capitalism in a World of "Difference,"' in P. Trifonas(eds.), *Worlds of Difference: Rethinking the Ethics of Global Education for the 21st Century,* Boulder, Colorado: Paradigm Publishing.

———— (forthcoming, 2009) 'Class-ifying Race: The "Compassionate" Racism of the Right and Why class Still Matters', in Z. Leonardo (ed.) *The Handbook of Education and Culture.* New York: Routledge.

Scott, A. (2006) 'Professor at Heart of a South American Dream'. *Lincolnshire Echo,* November 21.

Sheppard, R. (2006) 'The Assassinations of Malcom X and Martin Luther King, Jr'. *Fightback: The Marxist Voice of Labour and Youth,* June 14. Available HTTP: http://www.marxist.ca/content/view/161/50/ (accessed May 6, 2008).

Sivanandan, A. (1982) *A Different Hunger: Writings on Black Resistance,* London: Pluto Press.

———— (1985) 'RAT and the Degradation of Black Struggle'. *Race and Class,* 25 (4), pp. 1–33.

———— (1990) *Communities of Resistance: Writings on Black Struggles for Socialism,* London: Verso.

———— (2000) 'UK: Reclaiming the Struggle'. *Race and Class,* 42 (2), pp. 67–73.

———— (2001) 'Poverty Is the new Black'. *Race and Class* 43 (2), pp. 1–5.

Smith, D.G. (2003) 'On Enfraudening the Public Sphere, the Futility of Empire and the Future of Knowledge after "America"'. *Policy Futures in Education,* 1 (2), pp. 488–503. Available at: http://www.wwwords.co.uk/pdf/viewpdf.asp?j=pfie& vol=1&issue=3&year=2003&article=4_Smith_PFIE_1_3_ web&id=81.98.165.243 (accessed July 27, 2008).

Smith, D.J. (1977) *Racial Disadvantage in Britain,* Harmondsworth, UK: Penguin.

Solórzano, D.G. and T.J. Yosso (2007) 'Maintaining Social Justice Hopes within Academic Realities: A Freirean Approach to Critical Race/LatCrit Pedagogy', in Z. Leonardo (ed.) *Critical Pedagogy and Race,* Oxford: Blackwell.

Strand, S. (2007) *Minority Ethnic Pupils in the Longitudinal Study of Young People in England. DfES Research Report RR-002,* London: Department for Children Families and Schools (DCFS). Available HTTP: http//:www.dfes. gov.uk/research/data/uploadfiles/DCSF-RR002.pdf (accessed October 28, 2008).

Stovall, D. (2006) 'Forging Community in Race and Class: Critical Race Theory and the Quest for Social Justice in Education'. *Race, Ethnicity and Education,* 9 (3), pp. 243–259.

Suoranta, J., P. McLaren and N. Jaramillo (2000) 'Not Neo-Marxist, Not Post-Marxist, Not Marxian: In Defence of Marxist Cultural Critique in the Process of Becoming a Critical Citizen'.

Swann Report (1985), *Education for All: Report of the Committee of Inquiry into the Education of Children from Minority Ethnic Groups,* London: HMSO.

Talbot, A. (2008) 'Violent Attacks on Immigrants in South Africa'. World Socialist Web site (WSWS). Available HTTP: http://www.wsws.org/articles/2008/may2008/safr-m21.shtml (accessed May 23, 2008).

Taylor, M. (2006) 'University Suspends Lecturer in Racism Row Who Praised BNP'. *The Guardian,* March 24. Available HTTP: http://www.guardian.co.uk/uk/2006/mar/24/raceineducation.highereducation (accessed June 25, 2008).

Thane, P. (1982) *Foundations of the Welfare State,* London: Longman.

Thatcher, M. (1993) *The Downing Street Years,* London: HarperCollins.

Tomlinson, S. (1984), Home and School in Multicultural Britain, London: Batsford On the same page in the UNICEF.

—— (2005) *Education in a Post-Welfare Society* (2nd edition), Buckingham, UK: Open University Press.

Torres, R. and C. Ngin (1995) 'Racialized Boundaries, Class Relations, and Cultural Politics: The Asian-American and Latino Experience', in A. Darder (ed.) *Culture and Difference: Critical Perspectives on the Bicultural Experience in the United States,* Westport, CT: Bergin and Garvey.

Travis, A. (2003) 'Blunkett: Racism Tag Is Aiding Racists'. *The Guardian,* January 15. http://www.guardian.co.uk/politics/2003/jan/15/race.equality1 (accessed October 22, 2008).

Trotsky, L. (1909) *Why Marxists Oppose Individual Terrorism.* Available HTTP: www.marxists.org/archive/trotsky/works/1909/tia09.htm (accessed August 8, 2006).

Troyna, B. (1987) 'Antisexist/Antiracist Education—A False Dilemma: A Reply to Walkling and Brannigan'. *Journal of Moral Education,* 16 (1), January, pp. 60–65.

—— (1993) *Racism and Education,* Buckingham, UK: Open University Press.

Troyna, B. and B. Carrington (1990) *Education, Racism and Reform,* London: Routledge.

Tsosie, R. (2005–2006) 'Engaging the Spirit of Racial Healing within Critical Race Theory: An Exercise in Transformative Thought', *Michigan Journal of Race and Law,* 11 (21), pp. 21–49.

Tushnet, M. (1981) 'The Dilemmas of Liberal Constitutionalism'. *Ohio State Law Journal,* 42, pp. 411–426.

Unger, R.B. (1986) *The Critical Legal Studies Movement,* Cambridge, MA: Harvard University Press.

UNICEF (2007) UNICEF Report on Childhood in Industrialised Countries. News item, Febuary 14. Available HTTP: http://www.unicef.org.uk/press/news_detail.asp?news_id=890/ (accessed October 2, 2008).

U.S. Census Bureau (2006) *Poverty 2006.* Available HTTP: http://www.census.gov/hhes/www/poverty/poverty05/table5.html (accessed July 23, 2007).

U.S. Census Bureau (2007). Available HTTP: http://www.census.gov/ (accessed December 23, 2008).

Vail, L. (ed.) (1989) *The Creation of Tribalism in Southern Africa*, London and Berkeley, CA: Currey. Available HTTP: http://ark.cdlib.org/ark:/13030/ft158004rs/ (accessed May 23, 2008).

Van Auken, B. (2005) 'William Bennett's "hypothetical" on Racial Genocide: A Spreading Stench of Fascism'. World Socialist Web Site, October 3. Available HTTP: http://www.wsws.org/articles/2005/oct2005/benn-o03.shtml (accessed June 6, 2008).

—— (2008a) 'Obama and McCain on 9/11: "Unity" in Support of War and Repression'. World Socialist Web Site (WSWS) September 12. Available HTTP: http://www.wsws.org:80/articles/2008/sep2008/sept-s12.shtml (accessed September 12, 2008).

—— (2008b) 'Obama Continues Lurch to the Right on Iraq War and Militarism'. World Socialist Web Site, July 4, 2008. Available HTTP: http://www.wsws. org:80/articles/2008/jul2008/obam-j04.shtml (accessed July 4, 2008).

—— (2008c) 'Washington Demands Permanent Bases to Repress Iraqis, Launch New Middle East Wars'. World Socialist Web Site, June 6, 2008. Available HTTP: http://www.wsws.org:80/articles/2008/jun2008/iraq-j06.shtml (accessed June 6, 2008).

Vaught, S.E. and A.E. Castagno (2008) ' "I Don't Think I'm a Racist": Critical Race Theory, Teacher Attitudes, and Structural Racism'. *Race Ethnicity and Education*, 11 (2), July, pp. 95–113. Available HTTP: http://www.informaworld.com/ smpp/title~content=t713443511~db=all~tab=issueslist~branches=11-v11 (accessed October 28, 2008).

Venezuelanalysis.com (2007) 'Chavez Urges Party for All-Out Campaign for Venezuelan Constitutional Reform'. Available HTTP: http://www.venezuelanalysis. com/news/2569 (accessed August 15, 2008).

—— (2008) 'Chavez Nationalizes Bank of Venezuela on Last Day of Presidential Decree Period' http://www.venezuelanalysis.com:80/news/3687 (accessed October 22, 2008).

Verger, A. and X. Bonal (2008) Resistance to the GATS, In D. Hill (ed.) *Contesting Neoliberal Education: Public Resistance and Collective Advance*, London: New York: Routledge.

Vernel, P. and C. Carter (2008) 'Another Education is Possible'. *Socialist Review*, September. Available HTTP: http://rinf.com/alt-news/contributions/another-education-is-possible/4689/ (accessed October 20, 2008).

Viotti da Costa, E. (2001) 'New Publics, New Politics, New Histories: From Economic Reductionism to Cultural Reductionism—In Search of Dialectics', in G.M. Joseph (ed.) *Reclaiming the Political in Latin American History: Essays from the North*, Durham, NC: Duke University Press.

Virdee, S. (2009a) 'The Continuing Significance of "Race": Racism, Contentious Antiracist Politics and Labour Markets in Contemporary Capitalism', in A. Bloch and J. Solomos (eds.) *Race and Ethnicity in the 21st Century*, Basingstoke, UK: Palgrave Macmillan.

—— (2009b) 'Racism, Class and the Dialectics of Social Transformation', in P. Hill-Collins and J. Solomos (eds.) *Handbook of Race and Ethnic Studies*, London and New York: Sage.

Visram, R. (1986), *Ayahs, Lascars and Princes*, London: Pluto Press.

Waghorne, M. (2008) 'The Public Services International', In D. Hill (ed.) *Contesting Neoliberal Education: Public Resistance and Collective Advance*. London and New York: Routledge.

Walsh, D. (2008) 'Thousands of Iraqis Protest Agreement for Indefinite US Occupation'. World Socialist Web Site (WSWS) (2008) May 31. Available HTTP: http://www.wsws.org:80/articles/2008/may2008/iraq-m31.shtml (accessed May 31, 2008).

Walter, B. (1999) 'Inside and Outside the Pale: Diaspora Experiences of Irish Women', in P. Boyle and K. Halfacre (eds.) *Migration and Gender in the Developed World*, London: Routledge.

Walvin, J. (1973) *Black and White: The Negro and English Society 1555–1945*, London: Allen Lane.

Ward, P. (2005) 'Climate Change. A Large Scale Geophysical experiment? Global warming, capitalism and our future *Socialist Outlook*/08—Winter. Available HTTP: http://www.isg-fi.org.uk/spip.php?article314/ (accessed November 29, 2008).

———— (2006) 'Nuclear Juggernaut Moving into Top Gear'. *Socialist Outlook*, 9, Spring, pp. 12–13.

Watts, S. (1991) 'The Idiocy of American Studies: Poststructuralism, Language, and Politics in the Age of Self-Fulfillment'. *American Quarterly*, 43 (4), December, pp. 625–660.

Weaver, M. (2008a) 'British and US companies win Iraq oil contracts'. *The Guardian*. Available HTTP: http://www.guardian.co.uk/world/2008/jun/30/iraq.oil/ (accessed November 29, 2008).

———— (2008b) 'Live Blog: Barack Obama's Victory Sinks In', The Guardian. Available HTTP: http://www.guardian.co.uk/world/deadlineusa/2008/nov/05/barackobama (accessed December 31, 2008)

Weber, M. (c. 1915) [1947] *The Theory of Economic and Social Organizations*, New York: Free Press.

West, C. (1995) 'Foreword', in K. Crenshaw, N. Gotanda, G. Peller, and K. Thomas (eds.) *Critical Race Theory: The Key Writings that Formed the Movement*, New York: New Press.

Wetherell, M., M. Lafleche, and R. Berkeley (eds.) (2007) *Identity, Ethnic Diversity and Community Cohesion*, London: Sage.

Whitney Jr., W.T. (2005) 'Education Gets Huge Boost in Venezuela'. *People's Weekly World*. Available HTTP:http://www.pww.org/article/view/7279/1/275/ (accessed July 25, 2007).

Wikipedia (2008) 'Shouting Fire in a Crowded Theater'. Available HTTP: http://en.wikipedia.org/wiki/Shouting_fire_in_a_crowded_theater (accessed May 29, 2008).

Williams, R. (1980) *Problems in Materialism and Culture*, London: Verso.

Wilpert, G. (2007) 'Chavez Swears-In New Cabinet for "Venezuelan Path to Socialism'. Available HTTP: http://www.venezuelasolidarity.org/?q=node/32 (accessed May 4, 2007).

Woods, A. (1999) *Bolshevism—The Road to Revolution*, London: Wellred Books.

Woodard, C. (1986) 'Toward a "Super Liberal State" '. *The New York Times*, Sunday, November 23, Section 7, p. 27. Available HTTP: http://www.robertounger.com/cls.htm (accessed August 1, 2008).

World Socialist Web Site (WSWS) Editorial Board (2008a) 'Five Years after the Invasion of Iraq: A Debacle for US Imperialism', March 19. Available HTTP: http://www.wsws.org/articles/2008/mar2008/iwar-m19.shtml (accessed May 31, 2008).

———— (2008b) 'Five Years since the Invasion of Iraq: War, the Economic Crisis, and the 2008 Elections', April 17. Available HTTP: http://www.wsws.org/articles/2008/mar2008/isse-m29.shtml (accessed May 31, 2008).

Wrigley, T. and P. Hick (forthcoming, 2009) 'Promoting Equality: Pedagogy and Policy', in M. Cole (ed.) *Equality in Secondary School: Promoting Good Practice across the Curriculum*, London: Continuum.

Yarker, P. (2008) '*Learning Without Limits*—A Marxist Assessment: A Response to Mike Cole'. *Policy Futures in Education*, 6 (4), pp. 464–468.

Youdell, D., P. Aggleton, V. Hey, D. Reay et al. (2008) 'Youdell BBC's Whitewash of Industrial Decline'. *The Guardian,* March 13. Available HTTP: http://www. guardian.co.uk/media/2008/mar/13/bbc.television1 (accessed March 13, 2008).

Young, G. (2008) 'US Elections: Is Real Change Coming?' *Socialist Review* July/ August, pp. 10–13.

Younge, G. (2008) 'Obama's Army of Supporters Must Maintain Their Level of Activism'. *The Guardian*, November 10. Available HTTP: http://www.guardian. co.uk/commentisfree/2008/nov/10/barack-obama-supporters-campaign (accessed November 19, 2008).

Zarembka, P. (ed.) (2002) *Confronting 9-11, Ideologies of Race, and Eminent Economists,* Oxford: Elsevier Science.

Zephaniah, B. (2004) 'Rage of Empire'. *Socialist Review* (281), January, pp. 18–20.

Index